The United States and the Direct Broadcast Satellite

The United States and the Direct Broadcast Satellite

THE POLITICS OF INTERNATIONAL BROADCASTING IN SPACE

Sara Fletcher Luther

New York Oxford
OXFORD UNIVERSITY PRESS
1988

Oxford University Press

Oxford New York Toronto
Delhi Bombay Calcutta Madras Karachi
Petaling Jaya Singapore Hong Kong Tokyo
Nairobi Dar es Saalam Cape Town
Melbourne Auckland
and associated companies in
Berlin Ibadan

Published by Oxford University Press, Inc.
200 Madison Avenue, New York, New York 10016

Oxford is a registered trademark of Oxford University Press

Library of Congress Cataloging-in-Publication Data
Luther, Sara Fletcher.
The United States and the direct broadcast satellite: the politics of international broadcasting in space/Sara Fletcher Luther.
p. cm. Bibliography: p. Includes index.
ISBN 0-19-505138-6
1. Direct broadcast satellite television—Law and legislation.
2. Direct broadcast satellite television—Law and legislation—United States.
3. Space law. I. Title.
K4307.L88 1988 341.7′577—dc 19

Grateful acknowledgment is given to the following persons, journals, and publishers for permission to reprint excerpts from their publications: Beacon Press for Herbert I. Schiller, *The Mind Managers,* 1973. Copyright © 1973 by Herbert I. Schiller. Reprinted by permission of Beacon Press./Columbia Journal of Transnational Law Association, Inc. for *Columbia Journal of Transnational Law,* 1974, Vol. 13, No. 1. Copyright © 1974 by Columbia Journal of Transnational Law Association, Inc./George Braziller Inc. for C. Wright Mills, *Images of Man: The Classical Tradition in Sociological Thinking,* 1960. Reprinted by permission of George Braziller, Inc., New York./ George Washington University for "Toward the Free Flow of Information: Direct TV Broadcasting via Satellite," *The George Washington Journal of International Law and Economics,* Vol. 12, No. 2, 1979. Copyright © by the George Washington University. Originally published in Vol. 13, No. 2 of the *George Washington Journal of International Law and Economics.*/International Arts and Sciences Press for Herbert I. Schiller, *Communication and Cultural Domination,* 1976. Reprinted by permission of M. E. Sharpe, Inc., Armonk, New York 10504./McGraw-Hill Book Company for Malcolm M. Willey and Stuart A. Rice, *Communications Agencies and Social Life,* 1933. Material reproduced with permission of McGraw-Hill./Martinus Nijhoff Publishers for Edward McWhinney, *"Peaceful Coexistence" and Soviet-Western International Law,* 1964 and for Kathryn M. Queeney, *Direct Broadcast Satellites and the United Nations,* 1978. By permission of Martinus Nijhoff Publishers, Dordrecht, Holland./Monthly Review Foundation, Monthly Review Press for Frantz Fanon, *A Dying Colonialism,* 1965. Copyright © 1965 (English translation) by Monthly Review Press. Reprinted by permission of Monthly Review Foundation./New York University School of Law for M. S. Geston, "Direct Broadcasting from Satellite: The Case for Regulation," *Journal of International Law and Politics,* Spring 1970. Reprinted with the permission of the N.Y.U. Journal of International Law and Politics./Transaction, Inc. for "Direct Broadcast Satellites and Cultural Integrity" by Ithiel de Sola Pool, *Society,* September/October 1975. Copyright © 1975 by Transaction, Inc.

10 9 8 7 6 5 4 3 2 1

Printed in the United States of America on acid-free paper

Acknowledgments

Acknowledgment and thanks are given to my professors and fellow students at the City University of New York, especially Dr. Charles Winick, who originally encouraged me to undertake this study, and Dr. Michael E. Brown, who sustained and advised me throughout the project. I am grateful, too, to my colleague and friend Dr. George Snedeker, who so patiently listened and counseled. There is not space to enumerate the many scholars whose perceptive analyses of the role of mass communications in contemporary social formations provided background for the issues that are central to this book.

The resources of many libraries and the assistance of their staffs were important in carrying out this work. My appreciation is extended to the library of the Graduate Center of the City University of New York, the New York Public Library, the library of the University of Southern California, the library of Vassar College, Adriance Memorial Library (the public library of the City of Poughkeepsie), and the Hennepin County Public Library in Minneapolis.

Finally I wish to acknowledge the unflagging support and encouragement provided by my husband, Dr. John J. Neumaier.

Poughkeepsie, N.Y. S.F.L.
December 1987

Contents

The United States and the Direct Broadcast Satellite

1

The United States and the Direct Broadcast Satellite

Among the many sweeping changes effected in our lives by advances in science and technology is the possibility that information will be broadcast to millions of people over the world instantaneously. Historically, this kind of access to information was once the privilege of priests and kings—a privilege they were slowly required to share with ever-widening circles of participants. Today, almost everyone has access to information, via the incredible power of electromagnetic discoveries and their subsequent development in radio, television, and communication satellites positioned in outer space.

This startling but already mundane fact cannot be set forth without at the same time probing the multiple meanings of the term *information,* surely no neutral word. For example, by whom is information to be defined and organized? To whom is it to be directed? By whom is it to be received? At a broader level, the phenomenon of information exchange must be examined in relation to the central issues of our time, issues that loom over our daily lives with frightening persistence: How is humankind to survive the threat of world-consuming nuclear war? How is it to organize and direct itself toward reducing injustice and enhancing social and individual development?

Moreover, such questions, if they are to be posed at all, must be posed in the context of the conflicting ideologies that are shaping the contemporary world—that is, the deep and longstanding confrontation between capitalist forms of organization and socialist forms. There are many ways of perceiving this conflict, varying its definition, heightening its nuances, dwelling on or minimizing its contradictory aspects. And despite their differences, the two systems may be found to contain interesting parallel developments—for example, in the evolving status of women, the philosophy of education, the internal organization of industry, or the uses of ideology. Theoretically, such parallels can be seen as enriching our understanding of the complexity of social theory, highlighting commonalities rather than differences in modern industrial states, and perhaps permitting the clearing of new space for cooperative study and mutual understanding.

But in the larger arenas of world relationships—the forums of the United Nations, the campaigns for economic and political development in the so-called Third World, the ongoing national struggles over information access and cultural autonomy within world electronic and communication networks—the conflict between the two major systems and philosophies attains grave and, indeed, central importance.

Against this background has arisen the global controversy over the development of a technology called direct broadcast by satellite, or DBS. The possibility of broadcasting via outer space had been foreseen early in the twentieth century when scientists and science-fiction writers first began to speculate about human-made satellites. It came closer to reality with the emergence of rocketry during World War II and saw its initial demonstration with the first space satellite, the Soviet *Sputnik,* in 1957.

In the 1960s, the first satellites designed specifically for communications were launched. At first, they were positioned in elliptical orbits, circling the earth in ever-changing patterns of movement that required constant and difficult tracking by the sending and receiving earth stations. By 1965, the scientists and their gigantic rocket-launching apparatuses had accomplished the placement of communications satellites in the geostationary orbit (GSO). The GSO forms an arc directly over the equator at approximately 22,300 miles above the earth. Satellites launched into this orbit and fixed to travel at a velocity that matches that of the rotation of the earth appear to remain stationary in relation to the earth, thus permitting relatively permanent up-link and down-link signal transmission. The GSO orbit also permits global coverage: three such space communications satellites placed equidistantly in the GSO could transmit instantaneous radio or television signals to almost every area on earth. Moreover, since transmissions via satellite make use of ultrahigh-frequency wavebands (microwave), they are relatively free from the ionospheric limitations and static common to shortwave signals. They are also very difficult to jam.

While radio and television satellite transmissions have become a point of controversy, they represent but a tiny fraction of the total telecommunications transmitted by satellite. Most of the volume of today's space communications is represented by what is called point-to-point service, or Fixed Satellite Service (FSS). This means that the signals are transmitted by prior agreement between the sender and the intended receiver, often making use of complex routing plans and two-way circuits. FSS includes trunk telephony, television-program distribution, data transmission, and special services for business and manufacturing—for example, electronic funds transfers among banks, closed corporate data networks, and videoconference and videophone service. Use of FSS by different countries or by licensed entrepreneurs within countries has been accelerating during the past decade, partly because it offers enormous advantages for commercial and military telecommunications and partly because its development could proceed in an already existing regulatory environment, with accepted ground rules for intercorporate or interstate legal relationships.

The Broadcast Satellite Service (BSS) differs from the FSS in that the signal transmitted by satellite is intended by the sender to be received by earth stations for distribution to the general public, via individual or community receivers. During the late 1970s, variations of this kind of service developed rapidly in Western Europe, North America, Japan, and the Soviet Union. A specialized application of BSS has come to be designated as DBS—direct broadcast by satellite—in which the signal is intended for direct reception by the general public without an intermediary earth station. No DBS systems for television have been established—either domestic or transborder—although in 1986 in Western Europe, Great Britain, and Japan, plans for such services were nearing completion. In the United States, there is a kind of de facto DBS, as individuals set up their own dish antennas and capture satellite signals intended for reception by cable or network earth stations. This practice becomes international to the extent that neighbors in Canada and Mexico do the same.

In this study, the emphasis is on the controversy over DBS, although as the technology has advanced, a number of the regulations and issues involved are having repercussions for satellite communications in general, including FSS.

As space knowledge advanced in the 1960s, the potential of direct broadcast by satellite came to be of increasing interest to the various participants in world politics, including those engaged in ideological, developmental, and trade competitions. In the West, especially in the United States, the country with the most advanced space technology, DBS was assessed early as another possible channel for capital expansion, particularly into the markets of the Third World, which had already proved to be profitable outlets for the sale of American films, television programs, advertising, periodicals, and other cultural products. For other capitalist countries—Canada, the nations of Western Europe—DBS held a two-edged potential. If its uses were to be under the direction of United States interests, it could serve to accelerate U.S. cultural and commercial advances; but should the Europeans develop their own communications-satellite capability (as they subsequently did), DBS offered the possibility of creating a counterforce to U.S. penetration. In countries of the Third World, with their heritage of colonial exploitation and impoverishment, there was enthusiasm over the promise of overcoming underdevelopment with a technological leap to nationwide mass communication—in the service of such fundamental but still unattained goals as health care, literacy, nutrition, and agricultural advancement. But coupled with that enthusiasm was anxiety that the continuing concentration of space technology in the advanced countries was likely to work to the cultural and economic disadvantage of the smaller nations.

In the socialist countries, with the Soviet Union at their center, the new technology was viewed cautiously and in the context of past experience. Would the United States carry its ideological competition—Radio Free Europe, Radio Liberty (originally called Radio Liberation from Bolshevism), and Voice of America—from shortwave into outer space? How could socialist nations protect their sovereignty and internal autonomy from unwanted U.S. television penetration?

Out of this assortment of perspectives and concerns emerged an international drive to achieve a better understanding of the positive educational advances that could be achieved by utilizing DBS, a campaign spearheaded in the 1960s by UNESCO. At the same time, the Soviet Union led an international effort to persuade the United Nations to develop guidelines for international control of DBS. In 1972, the Soviet Union won General Assembly approval, by a vote of 101 to 1, of a proposal that the Committee on the Peaceful Uses of Outer Space (COPUOS) be instructed to develop guidelines to prohibit the sending of direct broadcast satellite television signals into a nation without its consent, to prohibit the broadcast of incitements to war, and to prepare such guidelines with a view to eventually embodying them in a treaty. The negative vote was cast by the United States, and in the succeeding years, it was able to forestall any favorable action by COPUOS on the development of such guidelines.

The U.S. position has been formulated in different ways, but essentially has rested on two arguments. First, on the United States Constitution, which is in this case interpreted as prohibiting government restraints, whether by treaty or by law, on U.S. citizens' reception or transmission of speech, whether internal or internationally. Second, on the perception of the United States as the conduit of truth to citizens of certain other nations whose right to receive information would be jeopardized should their government be allowed to refuse to accept incoming television signals.

Since 1972, while keeping the issue stalled in United Nations committee, the United States has moved forward with its own DBS technology, in conjunction with satellite communications development in general, by both private enterprise and the military, and by expansion of the U.S.–created bodies COMSAT and Intelsat.

The possibility of direct broadcast by satellite as a replacement for or competitor of existing domestic broadcasting arrangements, such as cablevision and the national radio and television networks, has raised significant regulatory and financial issues, both for the major communications corporations and for the Federal Communications Commission, which by law oversees the industry. Important as they are, these matters will not be addressed in this study. The emphasis will be on the complex political, social, and economic issues that have arisen over the possibility of direct broadcast by satellite across national boundaries into the domiciles of nationals of other countries. The development and possible regulation of this potential form of mass communication has continued to be a focus of intense international debate, with the United States, as the most powerful participant, at its center. This controversy and the role that the United States has played in its unfolding are to be examined.

One set of questions to be considered arises out of the possible uses to which DBS will be put. They are similar to the questions that were posed when mass broadcasting came into existence. For example, which of the many possibilities will become significant: educational tool, channel for entertainment, weapon for nation building, profit maker, social stabilizer, underground communicator? To what extent will the economic system within which the new technology is developed be reflected in the outcome?

Another set of questions turns on the history and present distribution of the world's technological and material wealth. Will the new technologies of communication be developed according to the existing patterns or reshaped to new demands and aspirations? Where and how is this possible contention manifesting itself?

Nationalism and national sovereignty have become concepts of transcendent importance. The unique capacity of DBS to go beyond the physical and ideological borders of nation-states obliges us to ask if and how it might penetrate or undermine twentieth-century sovereignties and their attendant social relationships and cultural and ethnic identities.

Central to these issues is the question of how and by whom this new mode of mass communication is to be controlled. As with other developments in the uses of outer space, the coming of broadcasting via satellite has required nation-states to recognize one another as co-inhabitants, even co-dependents, on the same planet, regardless of their differences. Their own self-interest makes cooperation imperative. In such a situation, how are competing national-security interests to be negotiated? What can be learned from earlier experience of enforced international cooperation in related areas—for example, when the telegraph and then the wireless first threatened to encroach on national sovereignty?

To understand the controversy over DBS and the role played in it by the United States, it will be useful to examine the history of communications development and the rise of international efforts to cooperate in controlling the use of the radio spectrum. The history reaches back to Paris in 1865, when European states began to work together to deal with international communication via telegraph. The delegates signed an international agreement that created the forerunner of what today is regarded as the oldest international specialized agency: the International Telecommunication Union (ITU). Official United States participation, or more accurately, nonparticipation, began at the much larger international telegraphy conference held in St. Petersburg in 1875. The complicated rationale for the United States' unwillingness to become party to treaty obligations concerning telegraphy policies and rates foreshadowed in many ways its present-day policy of opposition to controls over the international flow of DBS. Chapter 2 will explore the early phases of U.S. international communications history and trace the evolution of U.S. policy toward international regulation of radio. The rise of commercial radio in the United States after World War I and its belated domestic regulation, the continuing efforts of the League of Nations to regulate radio signals that crossed national borders, and the new phase of international cooperation that was the promise of the United Nations provide insights into how U.S. policy toward the issue of DBS has been shaped, by whom and with what goals.

For conceptual and methodological guidance, Chapter 3 turns to a study of the scholarship of American sociology in the field of mass communication. It reviews some of this literature, its early emergence, its development, and, especially, its consideration of the issues that arose with the sudden appearance and rapid growth of electronic mass communication. The examination is presented with a view to discovering the chief concerns and points of concentration of the researchers. It also seeks to bring out the connections that may have existed between the researchers and their institutional settings and the larger society in which they functioned, recognizing that profound changes were occurring in that society as it experienced economic depression, marked capitalist development, and the mobilization of all its resources for war.

Further, Chapter 3 seeks to determine the extent to which the emphases and methods of communications research in the 1930s were carried over to the postwar period of world-scale growth of American industry and influence. How did mass media scholars respond to new government and market initiatives and needs, and to what extent did their work seek or achieve objectivity? How did their scientific method function, and what were its essential assumptions? These lines of inquiry are conducted in order to determine how the accumulation of mass media studies and communications theory may be used to deepen understanding of the current DBS controversy and its ramifications.

One of the remarkable developments of our time has been the rapid rise of a complex network of international relations regularized by international treaties. The creation and growth of the United Nations and the emergence of its many specialized agencies have erected a structure for political cooperation among nations, and despite many setbacks, this new world political process has led to agreements of unprecedented scope and effectiveness. One of the central concerns of the United Nations has been the regulation of activities in outer space, culminating in 1967 with the Treaty on the Peaceful Uses of Outer Space, signed by the United States, the Soviet Union, and most other major powers. Concurrent with the forging of international treaty law in the United Nations, diplomats and authorities on international law have been giving sustained and concentrated attention to the form and content of international space law.

Chapter 4 provides an outline of the many views of what is thought to constitute international law, including the Soviet or Marxist understanding as compared with the Western interpretation. Within an international-space-law framework, it examines the difficult questions that appear to bear on the possible development and use of outer-space satellites to beam signals across national borders of sovereign earthly states. Does international law as it has evolved on earth, including the treaties and regulations of the International Telecommunication Union, apply in outer space? If international law and international space law do apply to a television signal relayed by space satellite, do they also apply to other signals relayed by space satellite? Does state sovereignty continue in, or extend to, outer space? Is there room for individual autonomy—free enterprise—in outer space? In the attempt to regulate the use of space in accordance with concept of international law, national sovereignty, and individual freedom, what recognition is given to military uses, already extensive and totally excluded from the jurisdiction of the International Telecommunication Union? What legal basis is there for the construction of a single global space communications system?

Exploration of such questions provides reminders of the tenuous nature of international law as a basis for problem resolution. Still, it is significant that experts in international space law from both East and West and the Third World are continuing to analyze and debate these issues. Some of their work has found its way into treaties and conventions. Much remains on the table, subject to power politics and international negotiation.

The central forum for the promulgation of international law is the United Nations, and it is there that the specifics of the legal debate over DBS have been laid out. Chapter 4 concludes with a summary and update of the years of meetings and discussions on DBS that have taken place there. It also discusses how the issue has taken on new dimensions in the form of the Third World's demand for a "new world information order," including a vast program of direct assistance, through the United Nations development plan, for modern telecommunications infrastructure and personnel that will ensure each country an autonomous role in world communication processes.

This study gives special attention to the role of the United States in the DBS debate. The United States has stood virtually alone in opposing the concept of a state's right to prevent being penetrated by an unwanted television signal transmitted by satellite. Chapter 5 turns to the specific arguments offered by the United States and by private interests within the United States to justify that position. Chief among them is the concept of "free flow of information." Its explication by American legal experts reveals the measure of its validity as well as some contradictions. Another central defense is said to be the mandate imposed by the First Amendment to the Constitution, which is interpreted as precluding limitations on broadcasting. Key analyses by recognized authorities in international law raise serious doubts about the limits and correctness of this argument. Other legal questions and the United States' interpretation of them are also explored in order to fully understand the U.S. position as, during the course of the long DBS debate, it has confronted opposing actions by the ITU, the nonaligned nations, and the equatorial countries. The chapter concludes with some of the speculation about two unspoken motives that may underlie the U.S. position. One is the U.S. fear that if television signals broadcast from outer space are restrained by international treaty, radio signals might also be regulated, thus legitimizing the Soviet Union's longstanding efforts to silence Radio Liberty and Radio Free Europe. The other is that regardless of any concerted action by other nations, the United States intends to proceed with direct broadcast by satellite across international borders as soon as it appears to be technically feasible and financially or geopolitically profitable.

Necessarily, there are troubling limitations in this kind of empirically oriented historical approach to a study of the problem of mass communication across national borders. It is not sufficient to recite facts and review history without relating them to the underlying forces of history and society, forces that are often central to the concentration over control of or participation in communications processes. As noted, Chapter 3 evaluates some of the methods of American social science and sociology and shows some of the inadequacies of traditional approaches. Chapter 6 examines another perspective, the method and substance of critical, or Marxist, approaches, as they may be applied to social communications on a mass scale. Finally, Chapter 7 provides an update on the rapidly developing field of global communications via satellite and offers some concluding observations.

2

The Political History of Transborder Broadcasting

The Early Development of Electromagnetic Communication

The rapid industrialization and accumulation of wealth that characterized western European nations and the United States in the latter half of the nineteenth century was both dependent on and a stimulus to improvements in communication. As land- and sea-transport technology expanded, so did the need for readily available information about the whereabouts and progress of trains and ships. The new capitalism needed banking and finance services, stable currencies, foreign-exchange arrangements, and far-reaching trading networks, none of which could exist or be expanded without ready access to information about market conditions, policies of partners or competitors in distant cities, threats to stability in those areas of the world that contained the raw materials, and the rise or well-being or fall of governments.

The industrialist's needs and the inventor's ability and willingness to try to meet them grew in tandem, reciprocally interacting. The milestones are well known. Two of the earliest were Samuel F. B. Morse's electromagnetic circuit breaker and his sparked message flashing along the telegraph wire from Baltimore to Washington in 1844: "What hath God Wrought?"[1] In 1851, the first undersea cable was laid across the English Channel. After enormous capital losses and technical difficulties, a transoceanic cable was completed across the Atlantic Ocean floor in 1866. The telephone came on the scene officially with the invention submitted by Alexander Graham Bell, who got to the U.S. Patent Office just ahead of Elisha Gray in February 1876. And in 1901, Guglielmo Marconi transmitted, in Morse code, the signal for *s* across the Atlantic, using not the cable but the "ether" itself—the atmospheric electric waves that the German physicist Heinrich Hertz had identified in 1887.

The inventors and their triumphs were part of a process of profound social and economic change. Significant aspects of that process may be perceived in the peculiar kinds of strife that engaged these entrepreneur-inventors—for example, their struggles to assert rights of ownership over their inventions, their roles in the larger trade and commercial contestations being waged among nations, and, especially, their ambiguous relationship with the states of which they were citizens. The variety of confrontations involving these issues—property rights, tension across national borders, and the relationship of the individual to the state—foreshadowed many of the complications that have come to be identified with the controversy over international broadcasting by artificial satellite. It is useful to consider these questions in some detail.

The tradition of private-property rights provided the framework and partial rationale for one of the fundamental elements of the Euro-American politicoeconomic system: the state's authority to issue patents. In the United States, the authority to issue patents had been assigned by the Constitution to the national government, building on the practice established centuries earlier in England.[2] The Constitution reads: "The Congress shall have power . . . to promote the progress of science and useful arts, by securing for limited times to authors and inventors the exclusive right to their respective writings and discoveries."[3] Under present law, the patent grant allows the patentee "the right to exclude others from making, using or selling the invention throughout the United States"[4] for seventeen years. Present law also requires that under certain conditions, permission from the government be obtained before an application for a U.S.–based patent can be filed in a foreign country. Significantly, this provision ensures a degree of state control over the "export" of technology, as does the authority vested in the Commissioner of Patents and Trademarks to "order an invention kept secret if publication would hurt the national safety or defense."[5] The patent process came to be of enormous importance in the history of telecommunication, as evidenced by the growth in intercorporate contests and court litigation over patent origins, patent sales, and patent rights throughout the period of development of radio and television.

The second issue—tension and competition among nation-states—should be understood in the context of the growing self-awareness that characterized nations in the nineteenth century. The process of nation formation and the growth of nationalism, although begun centuries earlier, were quickening with the spread of industrialization in western Europe.

Overlaid with a web of myths, goals, and motives, "nationalism" was defined and elaborated primarily through political initiatives, and it became the rationale for intense struggles. One essential condition for the development of effective nationhood was the control of internal communication, and as some nations expanded their political power and commercial enterprise in search of "empire" they sought to control communication on a world scale—a case in point being the rise of the British Empire.[6]

Great Britain was an early leader in the competition among nations to build industrial output and gain markets, and its successes both fed on and added to a rapid development and control of worldwide communication linkages. Britain's monopoly of undersea cables helped secure its supreme position in world politics and trade and served to tighten its hold over far-flung colonial offices, governors, and supply stations. Competitors, particularly American entrepreneurs and U.S. military interests, resented the profit taking and policy control exercised by British investors, as exemplified by the operations of a company like British Marconi. The rivalry and mutual suspicion between the United States and Great Britain continued for decades, presaging the international controversy and competition that has unfolded around the technologies of radio, including direct broadcast by space satellite.

The third issue—the role of the state in relation to the various power groups within it and, specifically, to entrepreneurs like Morse, Bell, and Marconi and to the companies they became associated with—raised questions that continue to affect the development of communication systems. The same Constitution that provided for the issuance of patents, also declared that "the Congress shall have power . . . to establish post-offices and post-roads,"[7] a grant of authority that placed these two kinds of services outside the scope of the marketplace and identified them as collective concerns of government. But what kind of postal service was this code of dots and dashes that Morse sent along a telegraph wire? Did it properly belong in the hands of government?[8] At first, Morse believed that it did, and he enumerated many possibilities for its use: as a means to verify bank accounts, apprehend deserters, provide needed national revenue, and "put agents of the state in instant contact with one another." But his lobbying efforts in Washington in 1844 on behalf of a nationalized telegraphy service were not successful; Congress decided otherwise, and, in time, Morse apparently changed his own views, pointing out that his business associates preferred private "arrangements."[9]

In other industrializing countries, the question of private versus public control of the telegraphy did not arise; the new telegraphy routinely became a part of existing postal services. That it did not in the United States may be seen as the beginning of an important pattern. Communications historian Keith Clark suggests that the decision was a conscious policy.[10] Congress had in 1843 appropriated $30,000 to let Morse demonstrate the usefulness of his invention. Another $8,000 was granted in 1845, with the disbursement being made through the Postmaster General. In 1846, Congress ordered that tolls from the telegraph line be used for the Post Office "in the same manner as other revenue from postage"; in 1847, "the experiment was abandoned." [11] Clark speculates that the "abandonment" was related to the recently acquired telegraphy interests of Repesentative F. O. G. Smith, who as chairman of the House Commerce Committee had helped push through the original $30,000 grant and who had subsequently left Congress to become a major shareholder in Morse's company.[12]

The profound changes following the Civil War did much to alter the economic direction and shape of industrial development in the United States.

Of particular importance was the rapid growth of the capitalist enterprise and the concurrent establishment of a dominant role for entrepreneurial interests in public decision making. In this context, national development came to be regarded as part of capitalist development. Telegraphy and, in the 1890s, telephony were developed in this way, as new arenas for the realization of profit, both for the inventor and for his back-up stock company. Little attention was given to any formalized concept of social or national need or to systematic public planning for communications development. Although in theory, government intervention, planning, financing, or regulation in the interest of the public was excluded from the capitalist development process, in fact, there were many such vital linkages between business and government. Businesses grew in step with such major government policies as land grants, issuance of patents, benign forms of regulation, direct subsidies in the form of military and naval contracts, and protection in the international arena when private U.S. commercial interests were in conflict with foreign governments or foreign entrepreneurs. For example, when investors sought backing for the enormously expensive work of laying transatlantic cables in the 1860s and thereafter, the governments of both Great Britain and the United States pledged financial support to their respective citizen entrepreneurs in the form of guarantees of future cable business.[13]

Another early and long-continuing policy was the decision by the United States not to ratify the International Telegraph Convention, which had been adopted by a number of European nations at a meeting in 1875 in St. Petersburg.[14] The decision was grounded on the view that to be a signatory implied acceptance of the concept of government regulation of telegraphy, which was repugnant to the private telegraph companies in the United States.[15] The International Telegraph Convention provided, among other things, that signatory states had the responsibility to ensure secrecy of telegraphic correspondence in their territory and had

> the right to stop the transmission of any private telegram which appears dangerous to the security of the state or which is contrary to the laws of the country, to public order, or to decency; and to suspend, either generally or only on certain lines, any or all international telegraphic service . . . subject to notification thereof to the other contracting governments.[16]

These provisions clearly implied a government's power to regulate telegraphy within its borders and between it and other nations.

While the refusal of the United States to ratify the convention may have indicated a rejection of regulation, there were also strong countervailing forces within the United States that supported a controlling role for the state in the management of communications infrastructure.[17] From the early days of cable telegraphy, the United States Navy had promoted the idea that the government was the rightful owner and operator of this new means of strengthening national security. Guarded support for the idea of the state as ultimate authority came even from commercial interests, jealous of the expanding reach of British and French investors who were seeking to secure their North

American cable terminals on U.S. territory without permitting reciprocity rights for American companies. In December 1875, President Grant issued a statement making it clear that any foreign submarine-cable company that sought to place cable terminals on United States land was subject to U.S. government control.[18] Portions of Grant's message follow:

> As these cable-telegraph lines connect separate States, there are questions as to their organization and control which probably can be best, if not solely settled by conventions between the respective States. In absence, however, of international conventions on the subject, municipal legislation may secure many points which appear to me important, if not indispensable for the protection of the public against the extortions which may result from a monopoly of the right of operating cable telegrams, or from a combination between several lines:
>
> I. No line should be allowed to land on the shores of the United States under the concession from another power which does not admit the right of any other line or lines, formed in the United States, to land and freely connect with and operate through its land lines.
>
> II. No line should be allowed to land on the shores of the United States which is not by treaty stipulation with the Government from whose shores it proceeds, or by prohibition in its charter or otherwise to the satisfaction of this Government, prohibited from consolidating or amalgamating with any other cable-telegraph line, or combining therewith for the purposes of regulating and maintaining the cost of telegraphing. . . .
>
> I present this subject to the earnest consideration of Congress. In the meantime, and unless Congress otherwise direct, I shall not oppose the landing of any telegraphic cable which complies with and assents to the points above enumerated, but will feel it my duty to prevent the landing of any which does not conform to the first and second points as stated, and which will not stipulate to concede to this Government the precedence in the transmission of its official messages, and will not enter into a satisfactory arrangement with regard to its charges.[19]

The Naval War College cited Grant's statement approvingly in a 1907 textbook and added a summary of an opinion by the Attorney General on the subject, which spelled out the commercial interests at issue:

> It is his (the President's) duty to see that in the exchange of comities among nations we get as much as we give. He ought not to stand by and permit a cable to land on our shores under concessions from a foreign power which does not permit our cables to land on its shores and enjoy there facilities equal to those accorded its cable here.

The attorney general's opinion also identified the security interests:

> A submarine cable is of inestimable service to the government in communicating with its officers in the diplomatic and consular service, and in the Army and Navy when abroad. The President should therefore demand that the Government have precedence in the use of the line, and this was done by President Grant.

Finally, the attorney general's opinion carefully circumscribed the basis of the president's authority: "The President's authority to control the landing of a foreign cable does not flow from his right to permit it in the sense of granting a franchise, but from his power to prohibit it should he deem it an enroachment on our rights or prejudicial to our interests."[20]

The Naval War College text illustrates the tension between the antiregulatory view, exemplified by the American telegraph companies in their opposition to the International Telegraph Convention, and the idea of the government as ultimate owner, operator, and controller, put forward by the navy for security reasons and occasionally by commercial interests when they needed protection from foreign competition. It is a tension that has characterized the relationship between the state and the dominant forces of the private economy since the beginnings of capitalism and industrialization in the United States. In the event, the issue of public ownership of telegraphy versus operation for private gain faded from view when telegraphy became more or less eclipsed by other forms of information transmission that arose with the technology of radio. Before that happened, a number of presidents, including Grant, Hayes, McKinley, and Theodore Roosevelt, had proposed that "electric mediums of communication should have the same relation to the government as have the mails."[21] Postmasters general repeatedly concurred in that view. Postmaster General Wanamaker was about to put telegraphy under government control, historian Keith Clark writes, but was stopped by an opinion delivered by acting Attorney General William Howard Taft, who found authority in the Constitution and the statutes for "transmitting telegrams, placing them in postal custody and delivering them" but none for the "employment or establishment of telegraphic appliances for the general transmission of correspondence."[22] The opinion must have been well received by the telegraph companies. In any case, it ended a debate that had lasted for some forty years. Telegraphy remained a private profit-making enterprise, and the United States government continued to remain outside the International Telegraph Union until 1932.[23] In that year, the International Telegraph Union was merged with the International Radiotelegraph Union, which had had its beginnings at the radio conferences held between 1903 and 1906, as will be discussed later.

The long controversy over the relative autonomy of the telegraphy industry in the United States has not been critically examined by communications historians. The few references that exist seem to treat the issue as a natural process devoid of social or economic stresses. Keith Clark, writing in 1930, summed it up as almost an act of nature by noting that "in a telegraph world of diversities" (meaning the difference between nonprofit government ownership and profit-making private ownership), adjustment is achieved by two actions: the great majority of governments entering into an international convention that executes international regulation, and "the small minority of countries remaining outside the union [to] regulate their telegraph services in whatever degree their constitutions and public opinion permit."[24] In 1945, Llewellyn White and Robert D. Leigh, writing about historical patterns of ownership and operation in international communication, ignored the long controversy over telegraphy operations in the United States and simply stated the outcome: "In contrast to the United States, the governments of virtually all countries are directly involved in the business of furnishing telecommunications services."[25] They did explain the exceptions: "In Greece, Egypt, Portugal and most Latin-American countries neither the government nor national industry controls the telecommunications links with the outer world, enterprising foreigners . . . having taken over such services by long-term lease."[26] That the "enterprising foreigners" were the same corporations that in the United States had insisted on freedom from government regulatory interference went unmentioned.

In sum, it is apparent that the ascendancy of the private commercial and corporate interests in the United States provided a temporary resolution of the controversy over control of telegraphy. But with the development of radio technology—telegraph without wires—in a context of rapid industrial growth and imperialist expansion, the issue of international regulatory controls was to reappear in new ways.

The Beginnings of International Control of Radio

The growth of electronic communication in the twentieth century was markedly accelerated by the forces of industrialization and rising nationalism, the emergence of powerful states, and, especially the wars and war preparations that industrialization and nationalism engendered.

Between 1898 and 1905, four serious colonial wars occurred: the Spanish-American War, the Boer War, the Boxer Rebellion, and the Russo-Japanese War. The outcomes of the first three strengthened the Euro-American position in world politics and world-trade expansion in a way that required immediate and vastly improved linkages to faraway lands and peoples, as well as more advanced naval defensive and offensive capability.[27] The new technology of wireless telegraphy (also called radiotelegraphy) promised to meet these needs.

By 1903 the British navy had set up a large number of wireless telegraphy stations and placed radio equipment on many of its ships.[28] Owners of merchant ships were also beginning to protect their cargoes and personnel by adding wireless equipment. In the United States, the navy was placing wireless equipment on its war craft and planning a string of coastal wireless stations.

International rivalry, however, exacerbated by intercompany rivalry, hindered the development of radiotelegraphy's possibilities. The Marconi Wireless Telegraph Company of Great Britain had gained a dominant position that made it possible for the company to follow a policy of refusal to communicate with systems other than its own; the operator of a Marconi apparatus was prohibited from responding to radio signals emanating from a non-Marconi transmitter, a policy that had the effect of blocking the exchange of critically important information relating to safe passage of ships. While this policy appeared to ensure the British Marconi Company's primacy, it aroused the ire of competitors, especially the Germans, who sought a resolution of the problem by convening an international radiotelegraphy conference in Berlin in 1903. (The first session was held in 1903; the last, in 1906.)

The conference attacked the British Marconi Company head on. With the exception of the delegates from Great Britain and Italy, the conferees decided to recommend to their governments that ship-to-shore wireless telegraphy "be exchanged without distinction as to the systems of wireless telegraphy employed."[29] The final document, issued in 1906, was called the International Radiotelegraph Convention. It stipulated that states that signed the agreement would "undertake . . . to impose the observance of these provisions upon private enterprises" or "on board ships which carry their flag."[30] But with Great Britain refusing to sign, the problem remained unresolved. Besides, despite the leadership of the United States delegation in Berlin in promoting acceptance of the "obligatory response" clauses, as they were described by the *New York Times,*[31] and despite strong urging from the United States Navy, the Senate Foreign Relations Committee was not willing to recommend ratification. Private companies testified that the convention might stultify "the free inventive American spirit as it evoked new wonders from the radio art" and that it "did not correspond with the continuous policy of the United States of refusing to accept international fixation of rates." The senators decided that the convention's provisions were "unjust, impracticable, hampering, premature, academic, unconstitutional," and they pigeonholed it.[32]

But the British Marconi Company's monopoly of the flow of maritime messages, while a temporary advantage, was slowly eroded by conditions largely beyond its control. The advance of wireless companies based in other countries was a factor. The United States Navy converted to a German system, the Slaby-Arco.[33] Between 1907 and 1912, some 250 ships of American registry installed equipment purchased from the U.S.–based United Wireless Telegraph Company.[34] In 1910, the sixty-first Congress took a step that must have seriously undercut the British Marconi Company. It required all ships to be radio-equipped and specified that such equipment would not be legal unless "the company installing it shall contract in writing to exchange . . . messages with shore or ship stations using other systems of radio-communication."[35] Further, as radiotelegraphic communication improved, its uses had expanded beyond maritime applications, especially into the area of news transmission by the Associated Press, United Press, and the *New York Times*—the *Times* establishing its own radiotelegraph station on a ship in New York harbor in order to obtain early news from incoming ships.[36]

Thus by 1912, there had arisen so many new problems affecting international wireless transmissions that Great Britain itself called a follow-up conference to the Berlin gathering. Only signatories to the 1906 International Radiotelegraph Convention were eligible to attend the London meeting. Since both government and private interests in the United States were determined to be present, the Senate Foreign Relations Committee renewed consideration of the dormant 1906 document and recommended it in April, just days before the sinking of the *Titanic,* an event that demonstrated to the whole world the folly of nations and communications companies in their long and stubborn resistance to rational planning and use of the life-saving capabilities of wireless.

By its belated ratification of the 1906 convention, the United States became a full-fledged participant in international treaty-making as it applied to wireless telegraphy, thus breaking the long-established tradition set by the wired telegraph industry, which continued to keep the United Staes out of the International Telegraph Union. Factors affecting this reversal may have included the pro-treaty naval interests, which had been relatively undeveloped in 1875, and the fledgling status of the wireless industry, which was not yet concerned about protecting its economic interests. The United States sent a delegation of twelve to the London conference, claiming six votes for the United States proper and six more for U.S. possessions, including Alaska, Hawaii, the Philippines, Puerto Rico, and the Canal Zone.[37] Although the convention rules allowed "mother countries" this type of representation, the six extra U.S. delegates were turned away on a technicality: insufficient advance notice.[38] In the event, the issue of the elegibility of delegates paled under the impact of the *Titanic* disaster, which hung over the conference and affected every decision. Quick agreement was reached on the obligatory exchange of messages between different systems, both ship-to-shore and ship-to-ship.[39] Much that had been decided at Berlin in 1906 was reconfirmed, and although no permanent organization was formed, the working relationship among the signatories of the International Radiotelegraph Convention became known as the International Radiotelegraph Union. The already existing bureau of the International Telegraph Union in Berne was designated as the administrative body of the International Radiotelegraph Union. The completed treaty was soon ratified by the United States Senate, and the domestic legislation necessary to comply with both the 1906 and the 1912 conventions was embodied in the U.S. Radio Act of 1912. This act required all U.S.–based radio stations transmitting across state or national boundaries to register with the government.[40] It also restricted radio transmissions by permitting the use of only those portions of the radio spectrum agreed on at the Berlin meeting.

Not unexpectedly, resistance to this new form of government regulation affected the drafting and subsequent interpretation of the U.S. Radio Act of 1912. One requirement of the international agreement was that signatory states were to regulate radio within their boundaries, a position that found little support among U.S. policy makers. So although the Radio Act authorized radio transmission only if the operators had registered with the government and received a license, the lawmakers insisted that this was in no way to be understood as implying any discretion on the part of the licensing officer (the secretary of commerce and labor) to withhold a license. Congressional commentary stated that "it is compulsory with the Secretary of Commerce and Labor that upon application these licenses shall be issued."[41] Thus the appearance of regulation was accomplished, and real regulation was made meaningless.

While the beginnings of a system of international regulation of radio signals were now discernible, the events of world history were to significantly accelerate the process.

World War I and Its Aftermath

The growth of capitalist states and, within them, the proliferation of systems of radio communication were profoundly affected by the war of 1914 to 1918. As in the past, warfare and war preparation served as tremendous stimuli to scientific and technological development: the interminable patent litigation could be peremptorily suspended, permitting scientists and technicians to work cooperatively; research could be undertaken on a very large, planned, and centrally controlled basis, with ample government support; private use of the airwaves was prohibited, allowing unimpeded government use and large-scale planning; and dramatic new experiments were possible—for example, battlefield communication by radio, and the unexpectedly effective radio dissemination of Wilson's Fourteen Points. As a result, there were notable advances in the use of the electromagnetic-wave-transmission processes. At the same time, the war brought about significant changes in the power and influence of nation-states, changes that were reflected in the prolonged and complex postwar negotiations undertaken to reestablish and stabilize world trade and order. A review of the postwar politics of international control of radio and of the development of radio as a mass medium will provide a useful background to the present-day controversy over global broadcasting from space.

Among the dozens of issues dividing the delegates to the Paris Peace Conference was the question of how to deal with the valuable German transatlantic cables, which had been taken over early in the war, one by the British and one by the French. (The British had cut both of the German transatlantic cables the day after the war started,[42] evidencing not only their accurate intelligence service, but also their concern at the possibility of transmission of German propaganda to the Western Hemisphere.) There were tangled ramifications to the negotiations at Paris concerning radio, among them the hostility of U.S. interests to the British Empire's longstanding domination of world cable traffic, and the strong suspicions of American businessmen that their British rivals could and did again gain access to information about their plans when cable messages had to pass through foreign connection points.[43] It was the general belief among U.S. negotiators that the prewar cable system had "been built up in order to connect the old world commercial centers with world business" and that now was the time to develop "a new system with the United States as a center."[44] There was also a growing awareness among Americans that radio, if properly developed and controlled, might be used to undercut, if not destroy, the huge advantages of cable, hitherto so valuable to the Europeans, especially the British.

Evolving out of these considerations came a U.S. policy that favored returning the captured cables to the Germans: "A divided control is more easily shaken than a monopoly," explained historian Ray Stannard Baker in his 1922 account of the peace negotiations.[45] This proved wholly unacceptable to England and France, and the situation soon took a new turn when President Wilson, counseled by Walter S. Rogers, proposed that the cables be held jointly, under international control or trusteeship, and that "the Five Allied and Associated Powers shall call as soon as possible an International Congress to consider and report on all international aspects of telegraph, cable and radio communications with a view to providing the entire world with adequate communication facilities on a fair, equitable basis."[46] Wilson's recommendation for Allied control disappeared, but, in an action that was to significantly affect communications development in the postwar years, the negotiators did adopt the proposal to call an international congress.

A word should be said about these proposals of Walter Rogers. They are unusual in the history of electronic communications in the United States, which, as we have seen, were strongly oriented to private development for profit. Rogers, a lawyer, businessman, and soon-to-be publisher (of the *Washington Herald* in 1919 and 1920), developed a plan for Wilson's consideration that appeared to promote international cooperation while serving American newspaper interests by effectively undercutting British cable power. High rates for the transmission of news dispatches over transoceanic cables were a concern of the owners of newspapers. Rogers proposed that the United States and other countries join together to internationalize radio, a new technology that "goes beyond the possibilities of private enterprise" and "the interest of any one nation."[47] He told Wilson that this would help promote understanding of the president's cherished League of Nations by increasing the flow of news about the league; another benefit would be that "a comprehensive plan can be devised for transmitting commercial press messages at low, uniform rates." An important advantage to be realized by creating this internationally controlled noncommercial radio operation would be the effect on cable companies, which would thereby be introduced to "fair competition," thus affording them "the regulation now so lacking over cable rates and services."[48] Rogers argued that this would not injure the cable companies because it was better than the alternatives, which he perceived as being either "going into the radio field" themselves, or "withstanding a disorganized and possibly demoralizing competition."[49]

The struggle between the cable companies and the growing radio interests complicated the intense rivalry between Britain and the United States for ownership and control of the existing and envisaged worldwide communications network so essential to the expansion of capital. Some historically important people were present in the halls of the Palace of Versailles, including Guglielmo Marconi, whose British Marconi Company was at the center of the struggle; Admiral W. H. G. Bullard of the United States Naval

Communications Service; and Bullard's superior, Secretary of the Navy Josephus Daniels, who, in keeping with navy tradition, was "a militant advocate of government ownership of wireless communications."[50] In the United States, although not far behind the scene, were the top men of the electrical-communication industry, most prominent among them Owen D. Young, head of the already powerful General Electic Company (GE). GE's impregnable position grew out of its ownership of the patents for the Alexanderson alternator system, a revolutionary device for extending radio communication that the British Marconi Company was determined to buy and that, if it did, would solidify its world communication supremacy in both cable and radio. From the earliest days of wireless (radio) emissions, there had been great technical difficulty in achieving adequate transmission of the human voice. Hertzian waves generated by the spark technology could transmit the dots and dashes familiar to wired telegraphy, and radiotelegraphy developed rapidly as continued improvements in spark-generation permitted better and faster signals. But radiotelephony—voice transmissions—required not sparks but the generation of a continuous wave of sound, an electromagnetic signal that could capture the nuances and modulations of speech. By 1918, the Alexanderson high-frequency alternator system had been shown to be capable of such radio transmissions. It was viewed as a technology that would outmode and replace the spark, in part because it was far more effective for radiotelegraphy as well as for radiotelephony.[51] The United States Navy was intent on preventing the transfer of this technological advantage to the British Marconi Company, and in April 1919, government negotiations led to an agreement by Young not to make the sale if, in exchange, he received the government's blessing for the formation of a GE subsidiary, the Radio Corporation of America (RCA), which would be given government contracts and patent advantages, and would assume peacetime responsibility for the navy's radio stations. This was accomplished, thus cutting off the British threat as well as the danger of government ownership and control of radio, as reportedly supported by President Wilson and Navy Secretary Daniels.[52] To assuage Daniels' concerns that RCA would become a monopoly in radio communications, Young is said to have argued that a monopoly is objectionable only if it controls a product or service, which RCA would not because it would be in competition with the communications enterprises of Great Britain.[53] Young also guaranteed that "control of the proposed corporation would rest in American hands" and that "a representative of the Navy Department would be named to the board of directors."[54] These long negotiations ran concurrently with the Paris peace talks, with the difficult months of Wilson's losing campaign for U.S. membership in the League of Nations, with Warren G. Harding's victory in the presidential election of 1920, and with the postwar reestablishment of the power and autonomy of business and industrial groups within the American political and economic structure. By March 1920, the government had returned all radio stations to their private owners; RCA had radiotelegraphed its first messages to England (at 17 cents per word, compared with the cable rate of 25 cents per word); and young David

Sarnoff had submitted a memo to his boss, Owen Young, anticipating the marketing of $75 million worth of "radio music boxes."[55] Understandably, the accounts of the genesis of RCA are recorded differently, depending on who is writing the history and under whose auspices. To the sympathetic radio historian Gleason Archer and, of course, to Young himself, who was to describe the founding of RCA many times for the popular press, it was an act of patriotic sacrifice. For the owners of General Electric and its new offspring RCA, it opened the way to incredible profit and power.[56]

International Growth of the Radio Business

Another major arena for the international politics of radio and cable during the early postwar years was South America, where both U.S. and European companies were engaged in keen competition to capture the Latin countries' communications systems and the markets that lay behind them. Elihu Root, Jr., corporation lawyer and former secretary of state, touched on the issues when he testified at a congressional hearing in May 1921:

> The war has now given this country an opportunity which may not occur again for another generation of getting back on the sea and resuming participation in the carrying trade, and the country has made a very great effort to take advantage of that opportunity. . . .
>
> The British are bound to make every effort to prevent that attempt of ours from succeeding, and they are past masters in the game of competition at sea. It is not a matter of ships alone. It is a matter of a good consular service, of coaling stations, of branch banks, and it is a matter of communications, and in that latter category the most important element is the cable. They understand that they have got to get to the ports where they trade with their own cables, and I do not blame their company for making the hardest fight possible to keep the Americans out of the east coast of the South American communication situation.[57]

The corporate negotiations over South America paralleled the talks at the Paris Peace Conference, with Woodrow Wilson's role being played by Owen Young of General Electric and the roles of the Allied leaders being performed by the heads of the British Marconi Company, the French Compagnie Universelle de Telégraphie et Téléphone sans Fil (which was controlled by Marconi),[58] and the German Telefunken Corporation. They gathered in Paris, in August 1921. Young wanted to prevent the British from gaining a communications monopoly in South America, and to that end, he proposed that the four corporations pool their existing South American holdings as well as their patents and proceed on a share-and-share-alike basis "to cooperate in the establishment and operation of such a number of such stations in South American Republics as may be necessary for the handling of the external communication in the best and most efficient manner."[59] The negotiations were difficult. As Young told it later, he won his consortium only by threatening to make a separate pact with the German Telefunken Corporation, to which the British and French were strongly opposed. One sympathetic historian asserted in his account that the consortium's provisions ensured against discrimination in the matter of "freedom of communication" because the agreement required that the chairman of the consortium be a U.S. citizen, selected by RCA, and that he be vested with veto power over any board decision.[60] This, the historian argued, carried "the principle of the Monroe doctrine into the field of communications," which he said was one of the original goals of Owen Young.[61] Neither this historian nor any other commented on the absence of any representatives from the South American countries in these negotiations over their "freedom of communication."

Few of the accounts of the negotiations reveal that at the same time that the big four negotiators created their consortium, they agreed on the formation of an international wireless trust (known as the Commercial Radio International Committee) that allowed for the use of one another's patents until 1945, thus providing each member undisputed control within its own country of the production of wireless apparatus.[62]

Significantly, from the standpoint of internationally agreed-on radio controls, the consortium was quick to adopt self-regulating principles, including specified wavelengths for long-distance international communications.[63] As for the companies' transmitting stations, it was agreed that communications would not be exchanged with any companies other than the members of the consortium, thus reviving the principle of nonreciprocity that the British Marconi Company had used so effectively in the early 1900s.

For the new entrepreneurs of radio, having navigated the stormy waters of the postwar reversion to private control of electromagnetic communications and the delicate business of protecting the "freedom" of the South American countries under the new communications regime of the consortium, there still remained the problem posed by the Paris peace negotiators' adoption of Wilson's call for an international congress to provide "the entire world with adequate communication facilities on a fair, equitable basis." Although Wilson and what he stood for had been replaced by Harding, planning sessions for such a convention were duly held in Washington and Paris, in 1920 and 1921. It will be recalled that the telegraph companies had succeeded in keeping the United States out of any international telegraph agreements. Although the Senate had finally, in 1912, ratified the convention that established an international radio organization, it had included a reservation to the effect that "it could take no action in the matter of tariffs [that is, rate-setting] as radio telegraphy was conducted in the U.S. by private enterprise."[64] This same objection was raised in 1920 and 1921 at the planning sessions for the projected international congress. American private companies opposed the idea of a treaty or convention that would appear to vest their government with power to set telegraphy rates. The historian John D. Tomlinson described the situation: "The United States was unwilling to impose [these provisions] on private companies because its Constitution did not 'permit.' The United States' proposal that this [exclusion of U.S. companies] be effected through the adoption of a separate protocol was more than the other powers were prepared to accept."[65]

In a footnote, Tomlinson adds: "The attitude of the American Companies against government regulation of rates is well known. It is possible that the election of November, 1920, had some influence on the position of the government in its support of the position of the private companies."[66] The upshot was that the international conference was not held.

What was at the heart of this position? Was it simply the question of government control versus private enterprise that many historians make it? But the British Marconi Company was a private company. Why did it not show the same hostility toward "government intervention" in its vast and profitable network of world communications? There are several possible explanations. First, the geography of Europe, with its many contiguous states, made cooperation in radio, cable, and telegraphy transmission an urgent necessity that could be met only by intergovernmental arrangements. A second reason for the difference might well have been the changing balance of capitalist power, as U.S. companies were catching up with and surpassing their British competitors; in their extremity, it is possible that the British industrialists were willing to gain strength by working in coalition with their government. Third, the differences in beliefs and traditions may have been a factor. The European state of the twentieth century carried in it a memory of the state of the royal monarchy, which perhaps provided it with a prestige not yet acquired by the national government of the United States. Were the contemporary British corporation directors mindful of that long period of successful mercantilism and early colonialism when the interests of commerce and the nation-state had been so effectively fused to accumulate "the wealth of nations"? Finally, how much of the difference was attributable to different levels of capitalist development? The ruling class of the mature, aging British Empire had had long experience with the dangers of government control and had learned how to manage and even how to profit from government power. But a still young and less experienced set of American corporate leaders perhaps feared the threats of populism, trust-busting, collective action by citizens, and the spread of ideas about social democracy and socialism, and were unwilling to make any opening that might strengthen a government over which they might not always have full control.

The Advance and Regulation of Radio Technology

In the years following World War I, radio was experiencing sudden and dramatic growth, facilitated partly by improvements in quality made possible by important wartime inventions; partly by skillful patent manipulation under the centralized corporate control of General Electric and its offspring, the Radio Corporation of America; partly by radio's enormous attraction as a device that almost magically seemed to enlarge people's ability to talk with one another; and partly by radio's emergence as an extremely profitable commodity for manufacturers and sellers of equipment.

The special qualities of radio (as opposed to those of cable telegraphy) had a bearing on the emerging efforts to regulate and stabilize its use, both internationally and in the United States. First, that radio was relatively new made it vulnerable to the possibility of government ownership. Ray Stannard Baker's comment represented the views of many: "In all cases [radio, cable, and overland telegraphy] private operation for profit is as serious an obstacle, particularly so far as the United States is concerned, to effective international cooperation as are national trade rivalries." He also took note of the view held by Walter Rogers, who "believed it would be easier to get hold of the still developing radio business than of the established situation in other fields."[67]

Further, radio's power as a direct propaganda tool had become apparent to wartime leaders, strategists, and manufacturers; telegraphy ensured secrecy and code protection and allowed for rapid transmission of information, but it had nowhere near the power of radio as a tool for influencing values, beliefs, and attitudes.

Another point of difference was in the capital outlay requirements and earning potential. Although wire telegraphy and, especially, undersea cables had exacted huge capital investment, they had served a paying clientele—of banks, businessmen, traders, colonial powers, press associations, and elite private users. Radio equipment was comparatively cheap and could be sold on a mass scale, permitting quick profits. (In these early years, little notice was given to another aspect of this difference—that mass use of radio represented a potential democratizing force that, unlike telegraphy, might affect the many who were not direct participants in the accelerating growth of industry. This will be considered in Chapters 3 and 6.)

Another significant difference was that of invisibility. Undersea cables and their landing terminals were openly identifiable, and their location frequently required bilateral negotiations between nations (recall President Grant's message), whereas radio waves can travel anywhere, unrestrained by governments.

Worth noting too was the vast change that had occurred in power relationships. The world telegraph network had been laid out around a central control point: London. Radio emerged at a time when such single-nation control was no longer possible.

These factors were to influence the postwar period of radio development, as Congress and the growing communications corporations took up the questions of how to shape, finance, and control radio; how to deal with the emerging international regulation; how to promote and profit from consumerism; and how to understand and mediate the social unrest that grew out of the economic convulsions of 1929.

Historically, 1927 proved to be a decisive year for U.S. corporate radio interests as they continued to defend their relative autonomy in the face of new pressures for regulation, arising both domestically and internationally. The political question of how radio broadcasting would be controlled reached a temporary resolution in February 1927, with the passage of the U.S. Radio Act. The industry owners were forced to accept the long-repugnant concept of government regulation largely because technical requirements of the radio spectrum made laissez-faire operation impracticable and unworkable. Competing stations, using the same frequencies, were obliterating one another's messages. From the early 1920s, the industry had attempted a kind of self-regulation, with Secretary of Commerce Herbert Hoover providing the semblance of a government arbiter. But in 1926, Hoover's approach had to be abandoned when the attorney general ruled that the U.S. Radio Act of 1912 made obligatory the issuance of a "license" to all applicants, thus ending any exercise of discretionary power by the secretary of commerce. The opinion was in line with two court decisions, made in 1925 and 1926. The chaos that resulted, as radio stations proliferated and competed with one another, brought President Calvin Coolidge to the rescue on December 7, 1926, with a proposal for legislation that would assign the function of deciding "who shall exercise the privilege of radio transmission" to a "board, to be assembled whenever action on such a question becomes necessary."[68] The matter had been under consideration long before Coolidge's intervention, and by February 1927, the proposal had become law. That the regulatory commission would be convened only as needed was an indication of the limited nature of the new regulation. Of further significance was the absence of any serious consideration of the possibility of radio's becoming a publicly owned and operated national service. Finally, the passage of this first piece of major radio legislation occurred only months before the meeting of a long-planned international conference on radio; clearly, the new law was intended to set the limits for any international agreement that might be shaped at the upcoming conference.

If the frontline for corporate radio interests in 1927 was the Congress and its actions on government regulation, the second crucial event of the year was the long-delayed postwar conference on international aspects of radio communication. We have seen how within the United States, new and accessible technology permitting the proliferation of radio transmission had forced enactment of a law that stipulated a modicum of government regulation. Similarly on the international level, the new technology—specifically, the emergence of shortwave broadcasting—was pressuring the nations and corporations to update their arrangements for regulating international radio communication. Many of the difficulties were directly related to the shortwave technology, "discovered" by amateur operators after they had been squeezed out of the other wave bands. Rapid progress was being made in both the United States and Great Britain in setting up shortwave transmitters and in learning how to use this new capability for sending signals worldwide at low cost, which had been thought could be accomplished only by using high-powered long-wave transmitters.

Another urgent reason for the conference was the appearance in February

1927 of the first transatlantic radiotelephonic service, transmitting voice sound rather than the dots and dashes still in use on the undersea cable. This advance in technology—which was engaging the attention of RCA, American Telephone and Telegraph (AT&T), and International Telephone and Telegraph (ITT)—repesented a substantial threat to the cable interests.[69] Another factor that helped bring on the international conference was the economic resurgence of Great Britain, France, and Germany and the growing competitive interest of the radio manufacturers in those countries.[70]

The United States was host to the international conference, which opened in October 1927 at the headquarters of the United States Chamber of Commerce with an address by President Coolidge.[71] Planning had been under way for two years, especially by groups and interests in the United States. As evidence of the close working relationship between the government and the radio industry, the official U.S. delegation included Owen D. Young, president of GE. Conference rules also permitted the attendance of private-company representatives in an advisory capacity, and there were scores of them.[72] Interested groups of observers held meetings in Washington concurrently with the meetings of the international convention, including the radio division of the National Electrical Manufacturers Association, and the International Association of Wireless Scientists. At congressional hearings held in 1928, Assistant Secretary of State William R. Castle said, in his description of the 1927 International Radio Convention, that "the private companies practically wrote it."[73]

Delegates to the conference came from eighty countries, and, predictably, the chief difficulty was that the companies in the United States did not want a document that would further broaden the appearance or reality of government control over them, while the delegations from most of the other countries viewed state control as necessary in order to maximize equitable use of the medium. Although the telegraph companies had long been successful in keeping the United States from signing the International Telegraph Convention, they feared the possibility of a merging of international telegraphy regulation with international radio regulation and wanted to forestall whatever change in their status such a merger might mean. American radio and telephone companies temporarily buried their intercorporate differences in order to form a united front against any treaty obligation that would appear to mean public intervention in their operations. In their view, the U.S. Radio Act of 1927 mandated as much regulation as was needed; should their transmissions spill over national borders into Canada, for example, the problem could be dealt with by voluntary negotiation between broadcasting companies, they argued. Hoover no doubt spoke for them when he urged the delegates to the conference to adopt regulation but to "put no harness on freedom."[74]

As for the general need for a new international radio agreement, a parallel may be observed between the situation that confronted the delegates to the international conference and the situation that had existed in the United States before the passage of the U.S. Radio Act of 1927. As worldwide radio broadcasting had grown, stations that transmitted across national borders had, in accordance with the agreement signed in London in 1912, registered their use of a particular wavelength with the international secretariat of the International Radiotelegraph Union. But there was no mechanism for either assigning or withholding slots; it was a system of first-come, first-served. Thus the companies or states with capital and technology were preempting the limited spectrum space, to the disadvantage of smaller or less advanced countries. Furthermore, messages were often obliterating one another. Resolution of the problem called for international regulation, but the U.S. companies, which, in effect, controlled the convention, held out for the barest minimum. The United States put forward virtually the same position that had been unacceptable to other countries in the failed planning sessions of 1920 and 1921, described earlier in this chapter.[75]

Specifically, while the U.S. companies agreed to an international allocation of specified frequencies for specified services (marine, aviation, long-distance telephony, police, and so on), they retained the provision "that each country was left free to assign any wavelength to any service, as long as it did not cause any interference with any service of another country."[76] This extended the first-come, first-served approach and left the problem of interference, between signals emanating from the United States and signals emanating from other countries, to be negotiated bilaterally, if and when it happened.

Understandably, the Senate's ratification of the convention proceeded smoothly. "The State Department could say that everyone was for it, that the United States Chamber of Commerce and the telegraph companies had endorsed it, and Senator Borah, chairman of the Committee on Foreign Affairs, could state that the Committee had not received a single protest against the convention."[77] Joseph O. Chamberlain, writing in the *American Bar Association Journal,* regarded the convention as the international expression of the new national law, the U.S. Radio Act of 1927.[78]

That the carefully designed international regulations were unworkable is illustrated by a statement made by the British delegation to the next international radio conference, held in Madrid in 1932: "[under the Washington agreement, a broadcasting station may] begin functioning in a wave outside the authorized band, with low power and without producing interference; subsequently it may increase its power and when we complain of interference, we are told that the public is accustomed to this service, and that it cannot be abandoned."[79]

In the matter of radio regulation, the long-range pattern of radio-broadcasting development in the United States and of the relations of the United States with other countrires was set by the decisions made during the 1920s. Officially, the international convention in Washington was occupied primarily with the allocation of wavelengths (Wilson's call for an equitable worldwide system of communication having long since been set aside). But there was an underlying concern about the authority of the capitalist state and the possible erosion of that authority should trends move in the direction of more extreme forms of democracy. Significantly, the Soviet Union was excluded from the 1927 international radio conference. As a signatory of the 1912 London convention, it was eligible to attend, but U.S. authorities refused admission to the Soviet delegates on the grounds that the Soviet Union was not recognized by the United States. The Bolsheviks' vigorous embrace of radio as a tool for building socialism was another possible cause for unease. As the various potentials of radio communication had become apparent, including its attractiveness to millions of people and its elusiveness in the face of authority, there had developed several levels of concern about getting it adequately harnessed. Navy and army personnel were anxious to retain as much control as possible for military use. The leaders of the major communications companies—GE, RCA, Westinghouse, and AT&T—had different perspectives, depending on their major business emphasis (for example, AT&T was mindful of the danger that radio might pose to its huge long-line investments, whereas the radio companies feared AT&T's technical and size advantages and the danger of competition from a combined radiotelephone device), but they all agreed that popular control via government regulation would be dangerous to their companies' futures.

A third group, often alluded to in the official histories, was the amateurs. Although few in number (by the mid-1930s, there were 75,000 amateur transmitting stations in the world, 48,000 of which were in the United States),[80] the amateurs had long been a problem for the dominant interests, both military and industrial. Early in the 1920s, because of their troublesome interference with the major users, the amateurs in the United States had been consigned to a virtually unusable portion of the spectrum. They experimented with it and thus, almost accidentally, furthered the development of shortwave. During the deliberations in Washington in 1927, they came close to being banned entirely and were assured permanent recognition only as a result of intensive lobbying. Unlike the military and commercial interests, the amateurs were closely regulated: the U.S. Radio Act of 1927 required strict licensing of amateurs, including passing specified tests; and restrictions were placed on making political statements on the air and on communicating across national borders.[81]

As for the military use of the spectrum, this had been recognized in the earliest international allocations made by the International Radiotelegraph Convention: the military was specifically exempt from international regulation, thus allowing it, in effect, to use any of the spectrum that was not allocated to other services.[82]

As we have seen, the general commercial broadcasters were brought under a form of benign regulation by the U.S. Radio Act of 1927, which, in effect, ensured radio's already established status as a private enterprise operated primarily for the profit of its owners. As the dominant force in the Washington international conference in 1927, the private companies helped write an international agreement that accorded them special status to continue to develop their use of the spectrum without restraints from other nations that might experience signal interference. A further decision of the international conference, to treat the Western Hemisphere separately from the rest of the world and allow it to make separate international agreements, appeared to protect the profitable radio relationships that U.S.–based companies were constructing in Latin America. The new treaty did allow for negotiation should interference occur and contained a provision for compulsory arbitration, although the latter provision has never been used. Perhaps most significant is that by virtue of being embodied in an international treaty, the provisions took on the character of "international law," including the principle and practice of allocating certain wavelengths for certain purposes. And since the treaty was ratified by sovereign states, it established the working principle that states were the entities responsible for the control of the new technology of radio. The notable exception was the United States, where a hybrid form of control was realized, combining limited state regulation with privately operated radio business.

Two more international radio conferences were held before the world's major powers found themselves again at war: at Madrid in 1932 and Cairo in 1938. At Madrid, the conferees effected a merger of the two international communications agencies—the one dealing with telegraphy and the one dealing with radio, forming what came to be called the International Telecommunication Union (ITU).[83] The ITU's bureau in Berne became the official administrator of the regulatory agreements. Continuing special arrangements were made for those countries that held to the "private-enterprise" approach, thus exempting U.S.–based telegraph companies from rate regulation. The major contestation at Madrid was over allocation of slots in the crowded bands that the conferees in Washington had designated for medium-wave broadcasting: 150–1500 kilocycles.[84] A new difficulty at Madrid was the presence of the Soviet Union. Since its exclusion from the 1927 Washington conference, it had allocated places on the broadcast spectrum to its various services with little regard for the international agreement made in Washington. A third problem was the increasing power of broadcasting stations, which was rapidly overreaching the regulatory capacity of the ITU.[85]

The Cairo conference made few major changes in ITU regulations, but one action should be particularly noted: for the first time, the conferees adopted *compulsory* worldwide allocation of specific frequencies for a specific purpose. That purpose was intercontinental airplane flight, for which twenty-eight shortwave bands were reserved. The allocation was not for particular stations, but for any station operating on a specified intercontinental air route, regardless of nationality. This remarkable collective decision setting clear limits on state sovereignty and corporate power recalls the origin of international action to regulate radio at the Berlin conference of 1903 to 1906. At that time, too, the threatened safety of costly maritime navigation apparatus served as the spur to seek suspension of national sovereignty in the interest of international action to protect goods and lives.

During the years that the ITU was trying to cope with the new radio technology, another international body, the League of Nations, was writing a little-known chapter of history on the potential and control of radio messages between nations.[86] The story is relevant to today's continuing arguments over the development and content of mass communication across national borders. In 1931, in part as a result of pressure from European broadcasting organizations, the eleventh assembly of the league requested the International Institute of Intellectual Cooperation, one of its specialized committees, to study the educational aspects of broadcasting, including "all the international questions raised by the use of broadcasting in regard to international relations."[87] The institute made preliminary studies and called together a committee of experts in early 1933. Over the next several years, a draft convention evolved, and the Council of the league convoked the Intergovernmental Conference for the Adoption of a Convention Concerning the Use of Broadcasting in the Cause of Peace, to be held in Geneva in September 1936.

Dr. Arnold Raestad of Norway was president of the conference. His opening remarks outlined the intent of the draft convention: "it aims at reinforcing the rights of every state in relation to all other states in the matter of political broadcasting" and seeks to define "the rights and duties of the State in this particular field." He noted that "broadcasts have no material substance, and therefore cannot be stopped at frontiers; they can be directed towards any point in space; the political effects may be extensive and immediate; but they are not easy to foresee or to control or canalise at need."[88]

On the basis of these views, the draft convention proposed a kind of radio disarmament treaty, "frankly universalist" even if at first only a small number of states were to adhere. The goal was to get agreement on principles of law that "will provide every state with solid foundations for the conduct of its own contemporary foreign policy through broadcasting."[89]

Specifically, the document required its signers to agree to prohibit the kind of broadcasting into another state that would incite a population to acts incompatible with the internal order or security of that state, would constitute incitement to war, was harmful to good international understanding, or was known by the broadcaster to be false. It also bound the signatories to "promote a better knowledge of the civilization and the conditions of life among the various countries, as well as their contribution to the organization of peace."[90]

The final act, containing these recommendations, was signed by twenty-eight of the delegations represented at the conference and took effect on April 2, 1938, having been ratified by or acceded to by nine nations (Brazil, the United Kingdom, Denmark, France, India, Luxembourg, New Zealand, the Union of South Africa, and Australia).

Subsequent world history has shown how fragile were the hopes of the diplomats at Geneva as they sought to harness the growing war propaganda and the destabilizing uses of international broadcasting. However, as Tomlinson remarks, "That any convention of this character could have been concluded and put into effect during a period of continued political crises is in itself remarkable."[91] Few who are involved in the present debate over direct broadcast by satellite have referred to this attempt made fifty years ago to regulate broadcasting "in the cause of peace," although it has occasionally been cited in Soviet literature.[92]

Immense changes in the field of international radio and television broadcasting have occurred since the outbreak of World War II, although fundamental patterns remain. Some of the political and regulatory developments will be examined in later chapters as we look more closely at the controversy over international control of the direct broadcast satellite. First, however, we review the ways in which media of radio and, later, television have been assessed and analyzed by recognized social theorists and intellectual leaders who have helped shape the ideology and outlook of American society.

3

The Shaping of Mass Communications Theory

Radio and Social Stability

As radio came into wide use, both within and across national borders, it also became the topic of study by social theorists and other intellectuals seeking to understand the potential of this radical new mode of communicating with people on a mass basis. It had arrived at a time of turmoil, marked by deepening social and economic division and by a new awareness of differing ideologies. The disastrous world depression that followed the stock-market crash of 1929 brought millions onto the streets, demanding improvement in their lives. Franklin Roosevelt in the United States used radio to talk to people in a way that no public figure had ever been able to talk to them before. The specter of the Soviet Union was haunting the leaders and parliamentary bodies of the major capitalist nations. The persistence of its experiment with the idea of a "people's government"—and its effective use of radio to reach out to the masses—was attracting the interest, and sometimes the enthusiasm, of activists in other countries and of workers who were experiencing the hardship of unemployment. In Germany, the ideals of liberalism and social democracy had proved unequal to the ascendant fascism of Hitler and his supporters. There, too, radio was used as a powerful tool to weld national support for Hitler's supremacist plans.

It was in this framework that communications scholars, social scientists, and publicists—the ideologues of capitalist society—began to look more closely at the new medium. The body of theories and data about mass communications that took shape in the 1920s and 1930s provides a basis for understanding what confronts us today; it guided the textbooks and instruction of future leaders, shaped the direction of research, and provided a rationale for the policies of government and corporations in the societies that engendered it.

After World War I, one of the first to recognize the implications of the accelerating exchange of information was an English historian of science, George Norman Clark. Like others to come, he identified communications as a portion of "the world's accumulated capital, a part of those possessions which are used for the creation of further wealth." [1] The concept of mass communication had not yet come into use, and Clark developed his own terms: *idea (idea-l) communication,* meaning "exchanges between cultures"; and *material communication,* meaning "exchange via boats and trains." Clark's purpose was to reflect on "the part played by the means of communication in international relations." From the start, he was bound by his own culture (what he called the "present industrial civilization"), as is illustrated by the example he offered of the possible future effects of the wireless: "One might draw a convincing picture of the 'special correspondent' of the future transmitting from his aeroplane to an attentive audience in the Alhambra Music Hall the full record, in phonographic and coloured stereoscopic moving pictures, of a riot in Punjab, or an earthquake in South America." [2]

That it would be the British who would sit quietly in the music hall and the Punjabis who would riot seemed only natural to Clark. And what else of interest besides an earthquake could happen in South America? Clark's ethnocentrism repeats itself in the literature.

Despite his cultural chauvinism, Clark identified many significant issues in his book. He saw how the use of expanded communication could homogenize culture and break down tariff barriers. He also identified the complex relationship between nationalism and the increasing flow of information: feelings of national unity could be intensified, but at the same time, such unity could heighten hostility toward external forces. Clark also confronted the question of the role of the state in facilitating or hindering communications development. He found that when states promoted the opening up of commerce or intercourse with foreign countries, it was often "as mere auxiliaries of economic 'interests' within them." [3] Clark did not foresee the coming of multinational corporations; naïvely, and perhaps because of his observation of the intense rivalries between British and U.S.–based companies, he concluded that "there is no world in which frontiers are more important than in the business world." [4]

Finally, Clark's study evidenced another concern that was to underlie the work of later social theorists and sociologists: the Russian Revolution and the arrival on the world scene of a different kind of social system. He predicted an ever-intensifying race by nation-states—capitalist and Communist alike—for control of sources of supply, a race that could be affected favorably by improved international communications. "A new organization [the League of Nations] is to endeavour to solve both the old problems and the still greater new problems which must be solved unless the civilization we know is to be destroyed," he wrote hopefully. But his conclusion was gloomy: the gigantic machinery of communications is still "unknown or misunderstood or neglected or perverted," and in trying to use it, we proceed much "like savages, who break up a mariner's compass to make jewels for their hair and tips for their arrows." [5]

In the United States, as in England, the work of early communications theorists and researchers reflected their political and social milieu, including in the United States, such factors as the relative isolationism of U.S. foreign policy; the continuing ascendancy of private industry and commerce; the widespread hostility toward Communism and the government of the Soviet Union, and a measured awareness that new and not fully understood forces had been set to work as a result of the use of propaganda during the "War for Democracy." Not long after Clark's book was published, Walter Lippmann, American public-opinion theorist, added a quite different dimension to the discussion with his well-known essay "The World Outside and the Pictures in Our Heads."[6] Lippmann's concern was with the unbridgeable gap between people's views of things (events, actions, politicians, and such) and what really exists. He emphasized the importance of fictions and symbols in human communication and saw a triangular process of (1) the scene of action, (2) the human picture of it, and (3) the human response based on that picture. For Lippmann, what the human being does is based not on direct knowledge, but on pictures "made by himself or given to him."[7] His insight led him to make an analogy that was to be used again by later social scientists: he likened the psychoanalyst, who examines how the patient adjusts to a given environment, to the social analyst, who examines the human being's adjustment to this pictures-in-our-heads pseudoenvironment. Lippmann set out to do the latter; that is, to find "how the political environment is conceived and how it can be conceived more successfully,"[8] or, put another way, why the pictures in our heads "so often mislead."[9] His enumeration of distortion factors included (1) the limits placed on people's access to facts (whether through overt censorship, through isolation or not enough time or too small a vocabulary, or through fear of facing threatening facts—surely a broad mix); (2) inner prejudices; and (3) the "interests" of an individual.

In the end, Lippmann recommended that there be "an independent, expert organization for making the unseen facts intelligible to those who have to make the decisions."[10] He felt that newspapers (then the only mass medium) were not capable of performing this task and proposed that it be done by political science. Indeed, he believed that the "perplexities of government and industry are conspiring to give political science this enormous opportunity to enrich itself and to serve the public."[11]

Lippmann's emphasis on the professional experts and his faith in the possibility of a neutral assessment of the validity of unseen facts was in the tradition of Max Weber, who had traced out elements of this point of view in his essay "Science as a Vocation."[12] Significant, too, is Lippmann's delineation of the process of communication as reciprocal, with the message sender evoking an action response in the message receiver (by constructing a picture in his or her head), although Lippmann did not develop the possibilities of this notion. Almost as an addendum to his essay was Lippmann's speculation on the difference between the power of the visual image (as conveyed by film) and the power of the written word. He regarded film as a new authority over imagination: "On the screen the whole process of observing, describing, reporting, and then imagining, has been accomplished for you."[13]

Lippmann's speculations were suggestive of the directions that were to be taken by social-science and public-opinion analysis, but it was the early work of the American academician Harold D. Lasswell that began to set the goals for sociology and political science as these disciplines turned to the study of mass communications.

The Emerging Social Science of Mass Communications

Probably no other American social scientist of the twentieth century is more closely associated with the beginning and early development of studies in propaganda, persuasion, and mass communications (as these studies eventually came to be called) than Harold D. Lasswell. Lasswell's prolific writing career spanned more than fifty years. He wrote as a political scientist, and from his first publication, he showed himself to be a strong believer in the Weberian concept of the neutral scholar who could scientifically examine the workings of society and interpret them for the general citizenry, including the policy makers. The social scientist, he said, is not just curious, but also analytical; his or her task is to discover and report, not to philosophize and reform.[14]

Lasswell reached adulthood during the time of World War I and the social stress that followed it. His first major work, published in 1927, deals with propaganda technique in wartime and in modern society. Although most of the book reads like a technical handbook on how to propagandize people, illustrated with impartially presented examples drawn from propaganda efforts during the war, the text does not entirely conceal Lasswell's underlying concern about certain disturbing social trends:

> Certainly, there is reason for believing that the propagandist who works upon an industrialized people, is dealing with a more tense and mobile population than that which inhabits an agrarian state. Industrialism has apparently increased the danger from those *secret mines which are laid by repression*. The rhythm and clang of exacting machinery is no less characteristic of the industrial way of life, than the blazing array of billboards, window displays, movies, vaudevilles, and newspapers, which convey *abundant and baffling possibilities of personal realization* [italics added].[15]

Lasswell was implying that because of this new condition—that industrialized people would become aware of the "possibilities of personal realization"—it is all the more necessary to understand and master the use of propaganda, the "new hammer and anvil of social solidarity," the "new antidote to wilfulness."[16] "If the mass will be free of chains of iron, it must accept its chains of silver," he wrote, in what sounds almost like a threat. The "chains of iron" appear in this context to be an echo of *The Communist Manifesto's* mid-nineteenth-century exhortation to the "workers of the world." Are the "chains of silver" a metaphor for the restraints of democratic law under abundant capitalism? Lasswell set forth similar thoughts in an essay, "The Theory of Political Propaganda," still writing in his rather oblique style:

> The ever-present function of propaganda in modern life is in large measure attributable to the social disorganization which had been precipitated by the rapid advent of technological changes. Impersonality has supplanted personal loyalty to leaders [Weber's typology?]. Literacy and the physical channels of communication [radio?] have quickened the connection between those who rule and the ruled. Conventions have arisen which favor the ventilation of opinion and the taking of votes [polls? elections?]. Most of that which formerly could be done through violence and intimidation must now be done by argument and persuasion. Democracy has proclaimed the dictatorship of palaver [echoing *The Communist Manifesto's* "dictatorship of the proletariat"?], and the technique of dictating to the dictator is named propaganda [playing on Marx's question, "Who will educate the educator?" from *Theses on Feuerbach*].[17]

At the end of his remarkable 1927 study of propaganda, Lasswell seemed to suggest that many of the political concepts of contemporary society were a cover-up for something else:

> To illuminate the mechanisms of propaganda is to reveal the secret springs of social action, and to expose to the most searching criticism our prevailing dogmas of sovereignty, of democracy, of honesty, and of the sanctity of individual opinion. "Democracy (and indeed all society) is run by an unseen engineer" [quoting Anatole France].[18]

For the next several decades, Lasswell continued to examine social groups and how they have been or can be subjected to propaganda, at one point extending his studies to include psychoanalytic theory, perhaps searching for those "secret mines which are laid by repression." In "The Theory of Political Propaganda," he traced out his formula for the propaganda process (building on his already developed definition of politics: the process of deciding who gets what, when, and how)[19]: Who carries it out, for what purpose, for what duration, toward whom, and with what results? Lasswell's formulation was to stand unquestioned in most subsequent studies of mass communication in the United States, as was his general theory of social science, including his insistence on the validity of the neutral observer and his faith in the empirical method.

After publishing his book on war propaganda, Lasswell turned his method and his concern with social stability to a large-scale study of the effects of Communist party propaganda on unemployed workers in Chicago at the height of the Great Depression.[20] When first published, in 1935, the study categorized, counted, and described the various techniques used by the Communist party (songs, slogans, fund allocations, rallies, membership rosters, and so on) and concluded that "revolutionary activity varies directly with deprivation," although to ascertain its effectiveness would require further data. However, Lasswell's scientific stance as a neutral observer did not prevent him from offering an analytic description of four counterpropaganda forces that might limit the appeal of what he called "world revolutionary radical propaganda."[21] They were, first, the difficulty that agitators created for themselves by building on "unstable human material," "personalities with the deepest anxieties," and by neglecting recruitment of the employed in favor of attracting "a disproportionate number of persons of a type predisposed to play disruptive roles in organized political activity." [22] The second force was the idea of nationalism. Lasswell noted that the Communist party was making wide use of American symbols and practices, and he emphasized how this tactic could be undermined by drawing attention to the party's connection with the Soviet Union.[23] The third counterforce he found was the concept of "American Individualism," which when invoked tended to strengthen rank-and-file localism against centralism and outside influence. Fourth, he located what he called a possible paradox in the propagandizing efforts of the Communist party: as it sought to destabilize the established order, it could actually be seen to be dissipating insecurities and strengthening "defensive tendencies against what was conceived as the threat of Russian control of American life and bureaucratic control of the destinies of the proletariat itself." [24] His suggestions resembled those that were being developed at congressional hearings and by ideology makers in Washington to combat alien ways of thinking, a consonance that harmonized with Lasswell's view of policy research. He believed fervently that the skilled professional social scientist could and should advise the policy maker, and he saw himself as engaged in building a science of propaganda, which called for the development of "its practitioners, its professors, its teachers, its theories," a science on which governments would increasingly rely.[25]

In 1935, Lasswell and his colleagues, working with the Advisory Committee on Pressure Groups and Propaganda of the Social Science Research Council, compiled a comprehensive annotated bibliography on propaganda.[26] In his introduction to the bibliography, Lasswell broadened his definition of propaganda to encompass his whole political theory: "Propaganda, in common with other social processes, has its bearing upon the distribution of values, upon who gets what, when, and how in terms of class, skill, personality and attitude." [27] In keeping with the social-science method, he was, in effect, seeking to free the term *propaganda* from its pejorative war-related connotations; it was now to be understood as the neutral use of words and symbols for the manipulation of group attitudes on controversial issues.

Lasswell also asked and answered the question as to why, in the mid-1930s, there was such wide interest in the study of propaganda. Recognizing its uses in waging war, in attaining and preserving power, in forging nationalism, and in supporting religion and commerce, he gave particular emphasis to the connection between the study of propaganda and "periods of rapid social change." [28] "When history was recaptured for revolutionary purposes by Marx, tactical questions connected with propaganda put the problem of the connection between 'material' and 'ideological' changes in the very center of social scientific inquiry." [29]

In these early works, Lasswell made little comment on specific media for propaganda dissemination—for example, radio, which between 1927 and 1937 was experiencing such amazing growth and worldwide attention. However, other scholars and students of society were becoming aware of the possible political and social impact of the new technology of electronic broadcasting. We have already commented on Herbert Hoover's involvement in the early promotion and eventual regulation of radio, during his years as secretary of commerce. In December 1929, then completing his first year as president and confronted with the collapse of the stock market and, possibly, of the whole economy, Hoover appointed the President's Research Committee on Recent Social Trends "to survey social changes in this country in order to throw light on the emerging problems." [30] The Rockefeller Foundation provided funds;[31] experts were recruited from universities; and from 1930 to 1932, a broad range of aspects of social life, including radio, was closely examined. The report on radio by Malcolm Macdonald Willey and Stuart A. Rice was published in 1933. Mainly descriptive and statistical, it nevertheless can be read as a legitimation of the existing patterns of radio development and ownership. For example: "By means of broadcasting, America has been turned into a vast auditorium into all corners of which a single voice can carry with dramatic ease and clarity." [32] The image of an audience of millions passively listening to a skillful (professional?) speaker evidences no perception of other ways in which radio technology might have been used—for example, by labor unions, cooperatives, farm groups, colleges, listener associations, and ethnic groups. That radio had been used by such groups during its early days was remarked on, with a table showing the decline in the number of small stations owned by educational institutions and church groups and the striking growth of commercial broadcasting. The authors commented that "the natural play of competitive forces was bringing about changes in types of ownership." [33]

Speculating on the social and political importance of radio broadcasting, Willey and Rice identified its likely effect on regionalism, including the widening of listeners' horizons to include, via shortwave, "auditory participation in distant foreign events as they transpire." "Social leveling" seemed another possible consequence:

> It [radio] is the agency of mass impression which, in its utilization, is more free from racial and class discriminations. The negro who cannot enter the white man's church, college or theatre can hear on his radio the white man's sermons, lectures and entertainment with complete equality; social isolation is to this extent broken.[34]

A section entitled "Mass Impression and Social Control" cited the same kinds of concerns that had turned Lasswell to his study of propaganda—specifically, the possibility of "concerted response to common stimuli." "Social mechanisms now exist whereby it is possible to impress the people of an entire country simultaneously. Any use of these mechanisms inevitably involves far-reaching effects upon social attitudes and behavior." [35] In a passage frequently quoted in subsequent texts on communications, Willey and Rice concluded: "A communication system fraught with greater possibilities for evil or for good has never before existed on so vast a scale. It is at once terrifying and inspiring; terrifying, because of the possibilities it opens for the accomplishment of selfish ends; inspiring for its potentialities of social self-control." [36]

Although Lasswell and the members of the President's Research Committee on Recent Social Trends were attuned to the possibility of radio's arousing large hitherto-uncoordinated masses by means of propaganda, their writings show no interest in the evolving relationship between the state and radio, as manifested in the debates over the U.S. Radio Act of 1927 and the U.S. Communications Act of 1934 and at the international conventions held in 1927, 1932, and 1938.

Lasswell's study of Communist propaganda during the Depression had led him to identify, as we have seen, how a heightened Americanism could serve to counteract the themes of Communism if those themes could be successfully labeled as foreign. But the full implications of nationalism's relationship to the uses of radio broadcasting were left for others to examine. One who did so was Oscar W. Riegel, who, like Lasswell, was at the beginning of a long career of communications and propaganda analysis.

Tracing the growing conflict among nations, Riegel suggested that an important causal factor was radio. In the days of cable telegraphy's dominance, he said, 87 percent of the undersea cables connecting nations and continents had been controlled by Great Britain, the United States, France, and Italy. "A condition of stability in the control of communications, with the Anglo-Saxon countries dominant, might have been maintained at least temporarily had it not been for the sudden rise of a new form of communication, radio." [37] Despite this destabilizing effect, Riegel saw in radio a potential for peace and a way to transcend nationalism, although it appeared to him that the apparatus of radio was increasingly being made "servile to the demands of nationalism." [38] No doubt alluding to the U.S. Radio Act of 1927 and the proposals for more federal legislation to regulate broadcasting, he found that "even in the United States the rights of private enterprise in electrical communication are rapidly disintegrating and may shortly disappear." [39]

Riegel aimed his strongest criticism at the developments in Madrid in 1932, when the International Telecommunication Union (ITU) had been formed (merging the International Telegraph Union with the International Radiotelegraph Union). He cited several areas where U.S. interests had been jeopardized by or had confronted resistance from other countries. One was the matter of voting power. From the beginning of international regulation of radio, as we have seen, the major colonial powers, as well as the United States, were allotted extra votes so they could "represent" their colonies, although there was disagreement among them on how these allotments should be made. Another was the difficulty of allocating radio frequencies for specific purposes. Riegel alluded to the "threat of a European raid upon channels previously allocated to the United States." [40] A third problem area was the threat of government control implied by such activities as internationally agreed-on rate setting. Always a touchy problem, it was more or less resolved at the Madrid conference when the American telegraph companies managed to get themselves excluded from the jurisdiction of international rate rules of the newly constituted ITU. Riegel was particularly concerned about "how much direct control the various national governments would be permitted to exercise over communications by censorship and confiscation of messages and interruption of international radio and telephone communication."[41] Arguing that "the well-being of the press everywhere depends upon the free and speedy flow of news from all parts of the world," he wrote that this issue had been ignored at the Madrid conference, except by the delegates from Great Britain, who had been motivated by British banking interests, and the delegates from the United States, who had pressed the issue because of the dependence of "American newsgathering agencies" on "the free flow of news."[42]

Riegel despaired over the political and economic restraints that endangered such a "free flow" and doubted the ability of the "average person" to cope with the complex flood of information and propaganda from foreign nations.

At about the time that Riegel was writing his book on the problems and trends of international radio, another academician, Rudolf Arnheim, was analyzing the same phenomenon, from a European perspective. The contrast with the American point of view is marked, especially with regard to the function of the state: "Wireless, in accordance with the different character of its technique, proved from the day of its birth to be so obviously a monopoly instrument concerning the whole of the community that in most countries it immediately came under the direct influence of the state."[43]

Arnheim's book is concerned primarily with the aesthetic effects of and differences among radio, film, and television, but his examination of what he called "Wireless and the Nations" set out some important political points that had been neglected by Lasswell, Willey, Riegel, and the handful of other U.S. social scientists who were then interesting themselves in the international implications of radio broadcasting.

That sounds made in one place might "be simultaneously reproduced in as many and as far removed places as one wishes, by disrespectfully breaking through boundaries of class and country" was to Arnheim "a spiritual event of primary importance."[44] This was so, he argued, because in the twentieth century, old barriers had begun to come under fire; unavoidable decisions were pending. The distance that once separated individuals, fostering isolationism and reliance on self, was giving way to a new proximity, a condition that "demands collectivism" and "promotes assimilation."[45] Especially advantageous for the movement toward community and away from separateness and isolation was the advent of radio "without prejudice," an advantage that Arnheim felt would be further enhanced with the development of television. Sound waves, he pointed out (as have so many others), pass by customs officers, need no cable, penetrate all walls, and are difficult to jam. "They make the whole world instant participators in everything,"[46] and "the possibilities of keeping from a single country facts which the rest of the world are discussing, or of spreading lies about other countries, vanish, for the voice of enlightenment sounds from without."[47] Arnheim embraced the idea of listening as activity, interpretation, sifting: "The listener who hears in the evening within an hour the latest news from London, Paris, Berlin, Rome, Moscow, and Basel" can "build up an approximate opinion for himself." "Let us imagine a war in which every wireless listener remained in daily contact by television with the life in the interior of the other country," he wrote. What would become of the "distorted caricature," "the figure of the enemy," so necessary to the creation of "a war-spirit"?[48]

Arnheim's enthusiasm was tempered with some negatives, including his recognition that as of the mid-1930s, only the privileged had access to good wireless sets, although he made note that nowhere had there been any attempt to make wireless reception a privilege of certain classes, as might have happened had the invention been at the disposal of feudal states. Like Riegel, he was disheartened by the observation that while people could learn facts, they were not capable of forming conclusions and judgments. He was also aware of the bitter struggle among nations for access to the limited number of broadcasting frequencies, a struggle that was manifested in the various international radio conferences. Ineffective compromises had led to what we hear today, he wrote, voices sounding "loudly on top of one another," "an artistically forceful symbol of constant war in peace, of the deficiency of central authority."[49]

In his discussion of the relationship of radio to the state, Arnheim observed that since everyone potentially "can get everything that is transmitted, then conversely, everything that is transmitted must be arranged largely according to principles laid down by the government." Since radio is a "universal commodity," it is "natural for such a matter to be either guided or even controlled by the state, like traffic, light and water, and street repairs."[50] At the same time, he contrasted the uniformity of radio under state control with the diversity of radio in the liberal state, citing as an example of the latter an event that had taken place in the United States: "in July, 1932, the American Worker's Guild founded with a capital of $120,000 a 25 kilowatt transmitter which, independent of the National Broadcasting and the Columbia Company, built up a purely socialistic programme in which the negro workers were also conceded regular hours."[51] He was referring apparently to WEVD (named for Eugene V. Debs), which, in 1931, had succeeded in reversing the decision by the Federal Radio Commission to take away its license.[52]

Arnheim was not unwilling to consider the idea of diversity and liberal pluralism, theories that he found wanting. Why use the radio to disseminate endless points of view without an underlying philosophy, he asked. There were dangers hidden in the question. Certainly, a "uniform" wireless could evolve by authoritative edict (which he rejected); but he argued that it could also arise as "the spontaneous expression of a unified national culture." He called this "organic"[53] wireless, a phrase that suggests the theories of Emile Durkheim. Durkheim believed that as the division of labor becomes more extensive in an industrializing society, the need for social cohesion and solidarity intensifies because the shrinking sphere of work activity provides less fulfillment to workers, making them rely instead on the larger network of social relationships represented by society itself. Durkheim wrote: "In effect, on the one hand, each one depends as much more strictly on society as labor is more divided; and, on the other, the activity of each is as much more personal as it is more specialized."[54] Durkheim called this form of solidarity "organic" and believed that in modern society (*sociétés supérieures* was the term he used), it served to knit together all individuals in a common purpose. Arnheim's "organic" wireless appears to provide another mechanism for strengthening the social bond. With "organic" wireless, there would be no conflict among radio listeners as to what viewpoints they wished to hear; the antithesis between individual and community occurs only "when there is no real communal life in being." Arnheim added:

> If wireless serves a state whose politics are an expression of the wishes and needs of the entire people, one single line of policy, unbreakable at any point . . . there will be no more talk of "tendencies" and "propaganda," there will be neither demands nor condemnations, but everything that is said, done and desired will be, involuntarily and imperceptibly, the natural expression of the "commonwealth" and the consequent trend of life.[55]

Arnheim felt that radio moves toward this goal by its uniform program, which both presupposes and promotes such an organic community. Specifically, it effects a leveling of taste and education across class lines, bringing the essential and characteristic art and thought to all.

The contrasts between the views of the European Arnheim and those of the Americans Lasswell, Willey, and Riegel provide a basis for understanding the emerging American field of mass communications. Arnheim's (and Durkheim's) idea of contributing to a sense of world community was seldom manifest in the thinking of American social theorists, whose methodology was based on their acceptance of the given framework of values; they seldom questioned fundamental mainstream beliefs and were seemingly unaware of their own role as stabilizers and legitimizers. By the mid-1930s, they were beginning to accept research support from the radio industry itself, as commercial communication interests sought to shape the theoretical context in which the social scientists worked, and the questions they set for themselves and society.

Trends in U.S. Mass Communications Research

In the United States, as we have seen, the social perspective on radio broadcasting was significantly influenced by the patterns of capitalist organization, which had clearly established the commercial, profit-making role of radio. This influence showed itself in two ways. First, there was little interest in any critical examination of the fundamental assumptions that had been made about radio ownership and control. Second, there began to accumulate a body of empirical data and specialized research that concentrated on various aspects of the broadcasting activity. Lasswell's famous definition of politics, "who gets what, when, how," appropriately amended to apply to broadcasting, "who says what, in which channel, to whom, with what effect," [56] provided the guideline for this large aggregate of American scholarship, of which we will examine but a few key samples.

In 1935, Hadley Cantril and Gordon W. Allport, who were then just beginning to build their academic careers, published *The Psychology of Radio,* with "gratefully acknowledged" assistance from NBC and CBS. The work illustrates many characteristics of American academia's social-science approach to the processes and conditions of American society. As self-declared social psychologists, Cantril and Allport specifically excluded the larger context and history of the phenomenon under scrutiny: "Our interest is not that of the reformer, the political scientist, or the legislator concerned with the moral or legal aspects of the problem. We are observers interested in understanding the psychological mechanisms involved." [57] They simply asked why people listen to radio and what happens to them when they do, leaving it to researchers in other disciplines to explore the circumstances that led to the listening event, including social conditions, political conflicts, and economic structures.

Cantril and Allport's study set forth a further refinement of the themes of Willey and Rice's radio report for the President's Research Committee on Recent Social Trends, emphasizing the "democratizing" effect of radio broadcasting in the United States and observing that the radio voice, unlike previous modes of communications, creates "the impression of natural equality among men." [58] At the same time, it fosters conformity, homogeneity, even "mob spirit." [59] Furthermore, because of the way "it promotes the interpenetration of national cultures," it presses always toward internationalism, toward "universal democracy," although "dictators" and "nationalists" may pervert this "natural property." [60]

Cantril and Allport identified another quality that in subsequent writing is less openly described: "Radio is perhaps our chief potential bulwark of social solidarity. It stems the tide of disrupting influences . . . and tends to counteract disintegrative forces." [61]

Much of the book is devoted to reports of empirical research on such matters as the differences between listening as part of a real audience and listening at home (the former was found to be the more powerful force for achieving conformity), the effectiveness of print versus radio (the latter was found to be more likely to bring about concerted action), the methods for successful propagandizing (Lasswell's methods are repeated, with the addition of a comment on the particular susceptibility of children), the wide variations in listeners' tastes and habits, and a comparative study of the male radio voice and the female radio voice (listeners preferred the male). The authors' claim to an objective and scientific approach provided them with the appearance of weighing "both sides" of an issue, but they generally concluded by affirming the views of the broadcasters. For example, in the matter of alleged censorship, after citing numerous instances of internal censorship by the broadcast industry, they concluded that

> since no . . . minority group has been able or willing to present evidence of having been excluded from the privileges of broadcasting and since the sworn testimony [in Washington in October 1934] shows a large amount of time, absolutely free of censorship, given without charge to such groups, it is impossible to avoid the conclusion that in general the broadcasters have been more than liberal in their support of the American traditions of free speech.[62]

At the end of the book, Cantril and Allport underwent an astonishing metamorphosis and became citizens and radio listeners instead of social psychologists. In these new roles, they expressed misgivings about the fact that radio, "except in a few countries, has been allowed to grow up as a device to increase the profits of a few competing entrepreneurs or as an instrument to secure the status of dictatorial governments."[63] But their suggestion that radio be freed "from the dictatorship of private profits" presents itself as a meaningless appendix, given their limited and superficial analysis of the political and economic system within which radio had developed. For example, they set forth the view that broadcasting is but a derivative of the habits and attitudes of the majority of citizens, whose economic and political habits function automatically, "seldom questioned and virtually unconscious in their operation," in favor of "laissez-faire individualism" and of letting "competitive business alone."[64] The two social psychologists found that "radio is not only a faithful interpreter of these attitudes but likewise it subtly intensifies those attitudes that support the status quo in broadcasting" by discouraging expression of dissent to such attitudes. Despite this rare moment of recognition of the role of the broadcast industry in discouraging dissent, Cantril and Allport's general emphasis on the idea that the broadcast media in the United States are but a reflection of the listeners themselves ("They get what they want") was to grow into one of the central tenets of mass communications researchers.

In the mid-1930s, another social scientist, Paul Lazarsfeld, began his career in the United States as researcher and evaluator of broadcasting. Like Lasswell, he was to figure prominently in the evolution of a body of ideas and data about radio as a social phenomenon. With funding from the Rockefeller Foundation, his Office of Radio Research was attached first to Princeton University and later to Columbia University, where it was energized by Lazarsfeld's own abilities and quick recognition of the research opportunities presented by American business needs.[65] One of Lazarsfeld's early books is *Radio and the Printed Page: An Introduction to the Study of Radio and Its Role in the Communication of Ideas*. Work on the book began in 1937 as a result of a grant from the Rockefeller Foundation to study "the role played by radio for different groups of listeners." In the introduction, Lazarsfeld mentioned his close collaboration with Frank Stanton, research director of the Columbia Broadcasting System, who "guided the author [Lazarsfeld] in his first orientation in the field of American broadcasting."[66] As the title suggests, the purpose of the book was to establish the lines of research that were deemed to require attention—for example, whether radio reaches different "strata of the population" from that reached by print; the problem of "information" versus "education"; what causes people to listen; and the effects of radio on reading, or "who listens to what and why." [67] Absent from the list were proposals for research into the social and political context within which American radio broadcasting was embedded or into alternative organizational and ownership patterns, including radio systems established in other countries.

After 300 pages of tables, charts, data, and comparisons (for example, Table 26, page 188, shows "Comparison between Total Amount of Listening and Amount of Listening to Boake Carter [popular radio commentator] in Syracuse and Columbus"), Lazarsfeld speculated that radio was unlikely to have profound social consequences:

> Broadcasting is done in America today to sell merchandise; and most of the other possible effects of radio [he had listed these as decentralization of the population, strengthening of the prevailing political system, "standardization," and "mass formation"] become submerged in a strange kind of social mechanism which brings the commercial effect to its strongest expression. There are no sinister tendencies operative in radio; it works all by itself.[68]

Lazarsfeld's book remains a recognized classic in the development of communications studies in America, a field in which the idea that radio "works all by itself" is still a major part of the mythology.

Also during the late 1930s, the topic of broadcast communications was beginning to find its way into sociology textbooks, generally in relation to public-opinion studies. The term *mass communications* was not yet in use. It did not appear in Lazarsfeld's book, nor had the key word *mass* been used by any of the analysts already discussed. A separate study would be needed to ascertain the precise moment when the terms *mass media* and *mass communications* came into use in American sociology, a use that, in effect, legitimized the word *mass,* reclaiming it from its Marxist context of "the masses." It is possible that its emergence stems from Herbert Blumer's essay on the varying manifestations of large groups, "the classic modern statement of the concept." [69] Seeking to clarify the various terms and their deeper meanings, Blumer offered definitions of three kinds of collectivities: the mass, the crowd, and the public. The mass, he argued, is characterized by the heterogeneity of its members (coming from "all walks of life"), by their anonymity, by their being "physically separated from one another," and thus by their lack of the kind of "unity that marks the crowd." [70] The mass and mass behavior he saw as growing in size and importance because of conditions of "modern urban and industrial life," including radio, which "detached individuals from customary moorings and thrust them into a new and wider world." [71] Elsewhere in his monograph Blumer differentiated "propaganda" from "mass advertising," arguing that "propaganda" seeks to bring about "collective action," while "mass advertising" tries to influence "individual action." [72] It seems likely that the usefulness of these finely honed definitions in depoliticizing the widespread consumer brainwashing that radio was then accomplishing and in legitimizing the descriptive term *mass* helped bring about their diffusion into other academic and commercial studies. It was not until after World War II, however, that researchers used the term *mass* in relation to broadcast communications.[73]

During the war years, the rules of neutral scholarship were suspended. Men like Riegel, Lasswell, Lazarsfeld, and Cantril assumed research tasks directly related to the support of the nation in its war effort, analyzing and putting into practice radio and propaganda techniques that would stabilize society, contain dissidence, and help strengthen national unity against the enemy.

An important example of this form of policy counsel is an essay written in 1942 by Talcott Parsons, "Propaganda and Social Control." Already recognized as one of the foremost American social theorists, Parsons was seeking in his essay to provide guidance for policy makers and reinforcement for fellow academicians and social scientists. The paper may be read as an instruction program for propagandists: it advised them to concentrate on the critical points of control in the social system, which are usually found in moments of interaction—for example, in the individual's search for the approval of others; it proposed that they work through the institutionalized control mechanisms—for example, the universities—and use appeals that would reinforce people's attachment to the prevailing social system. It may also be read as an example of Parsons's theory of society, made clear in an extended analogy between a sick society and a doctor's sick patient. The doctor-psychiatrist is portrayed as the propagandist, working from the vantage point of an institutionalized, and hence more prestigious, framework (for example, science, scientific culture, rational and liberal learning, association with universities). He confronts the patient/society, who/which may be suffering from deviance or ideological distortion, perhaps as a result of one of the undesirable kinds of propaganda that undermine people's attachment to the system. Just as the doctor treats a sick patient, so the propagandist administers a counteractive reinforcement, performing a kind of "social psychotherapy" for his society.[74] Like the doctor, the propagandist must identify with "the integrative elements," must work with existing institutions, must be regarded as impersonal, and must accept the "historic institutionalized values and cultural traditions."[75]

Underlying Parsons's formula was his own deep faith in the institutions of the United States, including medicine, religion, the universities, government, and law (only business was left out of his enumeration), and his unswerving belief in the ability of the professional to lead, heal, and direct—in effect, to rule over the ordinary or nonprofessional human being. Weber and Lippmann had had a similar faith in the professional, as we have seen.

In 1941, Harold Lasswell also spoke out on the role of the "specialists in intelligence," as he termed social scientists like himself.[76] Discarding his earlier emphasis on "propaganda," Lasswell took up the subject of public opinion, concluding from his analysis that it was in ever-present danger of turning against the public interest, which he equated with justice and majority rule. To keep this turnabout from happening, he recommended public discussion, including a policy of "free trade in propaganda" (a phrase that foreshadowed some of the debate over "free flow"). Another way to protect the public interest was to pay attention to the community's social structure, Lasswell wrote, which meant providing "respectable" jobs for all and reducing the gap between rich and poor. Essential to a vital and positive "public opinion," he argued, was full opportunity for participation in the social system, and he proposed, among other things, that corporations assign shares of stock to such organizations as farmers' associations, professional groups, and labor unions.[77] But the condition to which he attached most importance was the service of the intellectuals: "Our specialists in intelligence—including all who specialize in truth, or clarity or vividness (the journalists)—are those upon whom democratic society must depend,"[78] and "it is upon them that we must rely for improving the social inheritance of the future."[79] By then midway in his writing and teaching career,[80] Lasswell seems to have been becoming aware of structural deficiencies in the society he knew. But the self-imposed limits of his social-science method, which sought to reduce political and social processes to empirical formulas, held in check any development of a critical theory. He continued to believe that no crisis was immune to technical resolution; answers were to be found in "technical arrangements to provide an ever-expanding investment and an ever growing national income."[81]

As has now been widely acknowledged, the war period brought tremendous change in the relationship between academia and government, the repercussions of which still deeply affect all major educational institutions and most of the recognized scholars associated with them. During World War I, as we have observed, American capitalism became ever more closely linked with the state, a symbiotic relationship that grew stronger as the interests of both state and commerce expanded on a world scale. World War II accelerated this process and added another dimension by structuring American academic research and scholarship to serve the needs of the state. After the war ended, the overt identification of scholarship with military-industrial needs diminished as social science reinstated its more traditional rules. However, as before, the intellectuals, including the "specialists in intelligence," as Lasswell called them, continued to be profoundly affected by the events of the war and the vast changes that it had brought about, a large number of which concerned the politics of international mass communication. Of direct importance were the significant technological advances in electronics and in the use of the electro-magnetic spectrum and, in the United States, the official recognition of the tremendous political power of radio, as evidenced by the establishment of government-sponsored broadcasting into foreign countries. Until the outbreak of World War II, the protectors of the commercialized character of radio had prevented radio's direct use by the government to speak across international borders, and it was only after much debate that United States government leaders established the Voice of America to carry on the kind of world broadcasting that other major powers had been engaged in for more than a decade.

Other important events were to deeply affect the consideration of mass communications issues and the ongoing studies of international propaganda and its dissemination. Among them, the following are of particular importance for this study: the opportunity, after the failure of the League of Nations, to construct a worldwide organization of states, the United Nations; the beginning of an unstoppable progression toward decolonization, bringing to an end the formal empires of the big powers; the experience of wartime partnership between the capitalist countries and the Soviet Union and its abrupt breakup in 1946 with the reemergence in the United States of deep antipathy toward the Soviet Union; and world recognition of the United States as the most powerful and richest contemporary power. In the next section, we will consider how the shaping of mass communications ideology proceeded under these circumstances.

Mass Communications Research in the Postwar Years

Until the 1960s, the ideas of Lasswell, Lazarsfeld, and Parsons continued to dominate communications sociology and research, augmented by the work of their students and younger colleagues as they rose to positions of influence in academic, government, and corporate circles.

Lazarsfeld, still based at Columbia, to which he helped attract substantial funding from the broadcast industry, sought to fix the paradigms for what he now identified as "communications research."[82] One problem that he identified in the introduction to *Communications Research: 1948–1949* was the tension that results when educated people hear and see only what is "provided in the main by a number of large-scale, fairly centralized communications industries."[83] But while offering his readers a number of studies on listenership and readership (comic-book influence, reading habits during a newspaper strike, magazine choices, the "deviant" listener as represented by the low-income person who votes Republican, or "the influential," who sways the opinions of others), he included nothing on the concentration of the "centralized communications industries." The only study in the book that explores communications in a systematic and historical way was by Alex Inkeles on radio in the Soviet Union.[84] This essay marked the beginning of what was to become a noteworthy trend in the literature on mass media published in the United States: although relatively unconcerned with such questions as how ideology in American mass communication and communications research was entwined with class tension and economic and political conflict, the experts showed remarkable competence and critical ability when they turned their attention to the relationship between ideology and media in the Soviet Union.[85]

The emphasis on studying how listeners in the United States were affected by radio and television, as contrasted with studying how the broadcast industry was organized and controlled, was further strengthened with the publication in 1949 of Joseph Klapper's book *The Effects of Mass Media,* an exhaustive review of the already voluminous "effects literature."[86] While Klapper's book served to give credibility and prestige to studies of "effects," as opposed to historical or critical studies, it had another message. The author repeatedly asserted that these much-studied "effects" should not be regarded as lasting forms of propagandistic manipulation. He identified what he perceived as the many variables that were at work on the viewer and listener—for example, group norms, viewers' predispositions, and those important figures of American mass communications research, the "opinion-influencers" (Lazarsfeld's "influentials."). Despite the durable and seemingly benign influence of these factors, Klapper noted that things could be different in times of social unrest. He offered a list of the kinds of techniques that might be used to accomplish maximum persuasiveness. Chief among these was what he called "canalization"; that is, building an appeal on the basis of viewers' known attitudes.[87] Canalization, if correctly used, could have an impact in spite of the other variables. Klapper speculated:

> Given the rumblings of serious social malcontent—or, in terms of our current orientation, given individuals with predispositions toward change, unstructured as the envisaged change may be—mass communications would appear to be capable of molding or canalizing the predispositions into specific channels and so producing an active revolutionary movement.[88]

But Klapper discounted the likelihood of this kind of canalization occurring, at least in the United States, since "the nature of commercial mass media in a free enterprise society," including its economic requirements, necessitated "sanctification of the status quo." [89] Thus he reassured his readers that if there were to be canalization, it would support the predispositions not of the malcontent, but of the contented.

Two other publications of the late 1940s helped set the stage for the post-war social science of mass communications: Lyman Bryson's *Communication of Ideas* and Wilbur Schramm's *Communication in Modern Society*. Bryson reiterated the view that the organization and function of broadcasting, or "mass communication," had "this inescapable character: it is a relation between a single point of diffusion and a great mass of audience"; thus concentrated controls "are the substantial elements of any possible system of mass communication."[90] The contributors to his book did not show much interest in going beyond this definition, although within it, they evidenced disquiet. Margaret Mead cited the disintegration of the institutionalized centers of responsibility and called on broadcasters to adopt a code of ethics. Lasswell, having restated and enlarged on his equation (who says what, in which channel, to whom, with what effect?), found that the problem of the concentration of power in the hands of the broadcasters was being countered by audience talkback, such as that being provided through "scientific methods of sampling" (an apparent reference to the Neilsen and similar commercial rating systems).[91] Another prominent contributor to Bryson's collection was Robert Merton, who, during World War II and immediately after, had turned his attention to mass communications studies and had, in 1946, published his empirical "effects" analysis of Kate Smith's spectacularly successful radio program promoting the sale of war bonds.[92] Merton's essay in the Bryson collection, which he co-authored with Lazarsfeld, was entitled "Mass Communication, Popular Taste and Organized Social Action." In it, the authors took up the alleged danger of the power wielded by interest groups, including the possibility of organized business "manipulating mass publics through propaganda in place of more direct means of control."[93] In 1946, the Federal Communications Commission issued a report, "Public Service Responsibility of Broadcast Licensees" in which it warned that some radio trends were "inimical to the public interest."[94] Popularly called The Blue Book, this report documented and criticized the large volume of radio soap operas and commercial advertising and the relative absence of educational and public-service broadcasting. It "elicited cries of protest from the broadcasting industry on the ground that freedom of speech was being abridged."[95] Lazarsfeld and Merton appear to have taken account of the FCC report while writing their essay. Although they concluded that business support of the media had the effect of affirming the status quo and resulted in an absence of any effort "to raise essential questions about the structure of society,"[96] Lazarsfeld and Merton argued that this did not mean that mass persuasion could readily alter "deeprooted attitudes and ego-involved behavior."[97] To bring that about, they argued, would require certain essential conditions, which are rarely present all at the same time—monopolization, canalization, and supplementation through face-to-face contact. Lazarsfeld and Merton felt that these conditions had come together in the 55,000 reading rooms and radio-listening clubs in the Soviet Union in 1933.[98] From their analysis, they concluded that "the media do not exhibit the degree of social power commonly attributed to them"; indeed, they serve "to cement the structure of our society."[99]

It is, like Klapper's, a puzzling conclusion that reflects the dilemma besetting American media sociologists of the postwar era. On the one hand, they delineated the enslaving and propagandizing power of radio and television in the Soviet Union, where it was fully "monopolized," "canalized," and "re-enforced," on the other hand, they depicted the same technology in the United States as relatively harmless and diffuse, an institution that reflected prevailing beliefs rather than creating or enforcing them. The paradox is resolved when understood in the context of the evolving symbiosis of state and corporation, including the ideology of free enterprise, the profitable practice of media commodification, and the sustained support of and subtle influence over the communications researchers themselves.

It is relevant here to bring in the framework of functional analysis identified with Robert K. Merton, in part because of his involvement in propaganda and communications research and especially because of his dominant role, along with Parsons, in the field of American sociological theory. Merton was troubled by the absence of a general theoretical framework, a theory of society to which one might relate empirical studies. He solved the difficulty by giving up on it. The social theories of the past—for example, those of Auguste Comte and Herbert Spencer—he found to be no longer useful; they provided no "very solid base on which to stand." [100] So he redefined the problem by abandoning comprehensive theory altogether and calling for his famous "theories of the middle range"; that is, "interconnected conceptions which are limited and modest in scope, rather than all-embracing and grandiose." [101] He thus effectively removed the central problems of social existence—war, class conflict, poverty, ideological oppression, racism, sexism, imperialism, colonialism, nationalism, socialism, capitalism—from the agenda of the sociologist. To reinforce the limitation, Merton set forth requirements ("paradigms") for these middle-range theories (which he also called qualitative analysis): middle-range theories must be designed to emulate "the logical . . . rigor of quantitative analysis." [102] They must be "standardizable," "testable by others," and codifiable.[103] Perhaps most essential to his total outlook, which he called functional analysis, was his insistence that the research process clearly separate the subjective (that is, the social existence of the observer, which might endanger his or her research objectivity) from the objective. The separation is accomplished by the careful identification and removal of what cannot be empirically proved. Despite many refinements and subtleties, including the elaboration of the difference between latent function and manifest function, a distinction which Merton felt held great promise for advancement in sociological theory, the effect of his new method of functional analysis was not only to remove fundamental social problems from the sociological agenda, but also to strengthen the legitimacy of empirical research on problems that were relatively free from ideological overtones. For the growing number of mass communication researchers, it served as reassuring legitimation; their analyses were no longer a search for the insignificant. The introduction of the concept of latent function in social research led to conclusions that showed that "social life is not as simple as it first seems." For example:

> For as long as people confine themselves to certain consequences (e.g. manifest consequences), it is comparatively simple for them to pass moral judgments upon the practice of belief in question. . . . But the perception of further (latent) consequences often complicates the picture. Problems of moral evaluation (which are not our immediate concern) and problems of social engineering (which are our concern) both take on the additional complexities usually involved in responsible social decisions.[104]

Merton's favorable comparison of the American sociology of knowledge with the European is significant as a further legitimation of his methods. He championed the Americans, particularly the contributions made by mass communication researchers. Their neglect of historical material was understandable because of the unavailability of data: "The empirical rigor of the American persuasion involves a self-denying ordinance in which significant longterm movements of ideas in relation to changes in social structures are pretty much abandoned as a feasible subject for study." [105] Merton cataloged the many research techniques developed and used by Lazarsfeld, Stanton,

Hovland, Lasswell, Leites, and others in their efforts to determine how propaganda or persuasion affects listeners. "The student of mass communications . . . had almost from the beginning been concerned primarily with the impact of mass media upon audiences."[106] Merton compared this with the approach of European sociologists, who emphasized the source, the structural determinants of thought: "The European asks, how does it come to be that these particular ideas appear at all." That American research should focus on the listening audience was explained by Merton as due to the requirements of the market and the military, "since, with propaganda as with advertisements, the sponsors want to know if these have attained their intended effects."[107] He cautiously posed the question of whether government and industry had "too closely harnessed" the communications research, "providing too little occasion for dealing with more nearly fundamental questions of social science,"[108] which he deemed to be "problems of social incentive and motivation, of informing and persuading large numbers of men of the purposes and ends which should be pursued, and of having them adopt the expeditious ways of moving toward those ends."[109]

In sum, the emphasis of social-science research in the new field of mass communications studies in the United States was primarily empirical, concentrating on "audience analysis." It was a line of inquiry and a methodology guided and encouraged by the broadcast corporations and the advertising businesses for several reasons. It deflected attention from the growing public uneasiness about the "power of the media," and it precluded any serious critique of the established pattern of private ownership of the media. It also provided important information about ways of achieving effective impact on consumers of radio and television output. Another institution that influenced postwar research trends in mass communications was, of course, the government, supplemented by the eleemosynary foundations. Grants from these sources helped establish university centers "for the study of public opinion and communication by the methods of social science."[110] Commenting on these centers in 1950, Bernard Berelson and Morris Janowitz wrote, "Together with the continuing activities of industry and government, they now represent a large scale research enterprise."[111]

As we have seen, it was an enterprise with rather carefully circumscribed limits that asked and answered research questions that could do little more than stabilize the existing political and economic arrangements of the postwar United States, including the ownership and management of the mass media.

New Directions: U.S. Mass Media Research Goes International

By the mid-1950s, the publications of American specialists in mass communication began to reflect the new political and economic directions of the postwar period—specifically, the intensifying Cold War and the decolonization of the Third World. There were comprehensive studies of the media of the Soviet Union and a first-time awareness of the mass communications systems of the nonindustrialized world, including the territories that were emerging from British and French control. Both trends were fostered by and, in many cases, directly initiated by the departments of State and Defense, private foundations, and, covertly, the CIA. Wilbur Schramm, who became an authority in this area of study, described it as sponsored research "into the symbolic battle for the minds of men and the friendship of nations." [112] A major tool for those embarking on this new kind of study was the updated edition of Lasswell, Casey, and Smith's annotated bibliography, first published in 1935 as *Propaganda and Promotional Activities*. It had been revised and reissued in 1946 as *Propaganda, Communication, and Public Opinion: A Comprehensive Reference Guide*.[113] It came out again in 1956, with a change in emphasis and with funding from the Rand Corporation (a research institute closely linked to the United States Air Force and the Defense Department), as *International Communication and Political Action: A Guide to the Literature*.[114] Bruce Lannes Smith, who had been a co-editor in 1935, was the principal editor of the 1956 version. His introduction stressed some of the points that he felt helped explain the expansion of communications research to include international communications. He cited, for example, the role of the United States government as sponsor of "worldwide propaganda on a large scale," [115] the rise of the foundations' interest (he cited the Hoover Institute; MIT's studies in international communications; and Princeton's Center of International Studies, especially its Appeals of Communism series), the growing emphasis of UNESCO on studies of international communication as a way of reducing tension, and the needs of business firms engaged in international trade. In effect, he called on communications researchers to turn their attention to the problem of "developing an integrated, non-authoritarian, international information system as an instrument of healthy socio-economic growth." [116] He found that "Free World research, which might have assisted somewhat in clarifying Free World public policy along these lines, has tended to leave the Communists a great deal of the initiative in announcing ideas about how to bring 'the good life' to the less-developed areas." [117] He continued:

> The field of "Free World development communication" and the training of personnel to take part in it would seem to be full of challenging possibilities. Certainly it is still wide open for inventive theorists, methodologists, and field workers, and for a leading school to develop a coherent plan of research.[118]

With the added stimulus of the Soviets' launching of *Sputnik* in 1957, the universities and their communications faculties responded to the challenge to enter the field of development communications studies. Communications research centers multiplied. Symposia were planned, and publications appeared, including studies of communication in Africa[119] and of the U.S. Information Service,[120] studies linking national development with communications development,[121] and analyses of the direct foreign-policy implications of international communications.[122] An example of the last is Walter

Phillips Davison's *International Political Communication,* published in 1965 under the auspices of the Council on Foreign Relations. Davison, as the unofficial dean of international communications studies, concentrated on the problem of presenting the American image, arguing that the United States should "build more purposefully on its traditions of free public discussion,"[123] champion freedom of information, adhere to the objectivity and truthfulness of the best private media, and open a two-way exchange with other nations. Davison felt that ideas directed at Communist nations or emerging countries should effectively advance U.S. foreign policy, a goal that called for better knowledge of the potential audiences and for closer partnership between the communications industry and government. He also recommended more coordination in the presentation of events that sometimes appeared to be contradictory—for example, when a disturbing picture might be created by reports of individual racial incidents rather than an accurate report of race relations in general.[124]

Hadley Cantril, who by the 1960s had become chair of the psychology department at Princeton University and had written many additional books and reports on propaganda, including *Soviet Leaders and Mastery over Man,* entered more directly into policy advising with a number of propaganda recommendations submitted confidentially to former CIA Director Allen Dulles in a memo dated March 21, 1962.[125] Among the themes that Cantril felt U.S. propaganda should repeat and repeat "in rifle shot fashion" were "that the Communist Party uses nations and states and the people who compose them as instruments or tools to carry out Party policies"; that the issue in the underdeveloped areas is "not one of 'communism' versus 'capitalism' but the dictatorship of the Communist Party versus the possibility of national development in a free and open way"; and "that while Soviet leaders talk about 'co-existence,' in reality they seem to be afraid of it. They object to open skies and to an open world." Cantril also urged Dulles to bring about the formulation of an ideological statement of purpose in order "to out-compete the Soviet Union in teaching people in underdeveloped areas what to want—what 'progress' and 'civilization' can and should refer to." He recommended a policy that would praise the Soviet people's advances—for example, in their standard of living—since the effect of such praise would be to further build the United States as a model and thus help continue pressure in the Soviet Union for more consumer goods.

Application of the kinds of policies recommended by Davison and Cantril, among others, became increasingly difficult as the emerging nations took their places on the world stage and as enormous advances in communication technology sharply increased the number of people in the world who could learn about current events. One may discern a similarity here with that earlier period of communications development within the United States when authorities grew anxious about the explosive growth of radio and the possible effect on millions of previously isolated citizens as they became exposed to a national communication network of an impact and power not yet known. Now the number of new listeners was in the billions, and data on what they might be listening to, and with what "effects," was not readily available for research purposes to U.S. communications scholars. Their techniques for content and audience analysis would be useful in assessing the new phenomenon, especially in the context of the perceived propaganda competition between the Soviet system and the U.S. system. But they needed a wider scope than the one to which they had accustomed themselves if they were to probe the relationship between mass media and economic development in the Third World.

This need led various theorists to define the problem as one of modernization, or of transitional societies suddenly confronted by the phenomenon of "leapfrogging," referring to what was regarded as a sudden leap by the Third World into the twentieth century without having passed through the arduous 500-year progression of the West from Gutenberg to the present informed state of civilization. The difficulty of applying North American social-science concepts and approaches to the Third World was compounded by uncertainty over what kind of base the new countries would use to construct their societies and their communications networks. Was it to be "free enterprise," as in the United States, regulated by a benign governmental agency like the FCC? This was the pattern to which the media sociologists had become accustomed and which their research had generally accepted as a given. Or was it the time to again ask old questions, such as those touched on by Cantril and Allport in 1935, by Lazarsfeld and Merton in their reflections on the problem of centralized corporate control of mass communications in the postwar era, or by the more recent statement of the Hutchins Report that "the present advertising-commercial control of programming in the domestic field is quite unsuitable in the international field."[126]

Most of the communications researchers who took up the study of media in the developing world did not raise the issue of how the mass communications network was concentrated and controlled in the United States, although they accepted First Amendment freedoms as fundamental: "A free, independent, autonomous, and preferably non-partisan press is an essential prerequisite for democratic government."[127] Given the remoteness of this ideal in the impoverished new nations, not to mention the difficulties of explaining the acknowledged oligopolistic patterns of media ownership and control in the United States—including the U.S.–based, worldwide newsgathering corporations; the conglomerate holdings of radio and television networks; the transnational expansion of such corporate communications developers as AT&T, Ford, ITT, Hughes, and RCA; and the vast government-owned and government-operated Voice of America—it was understandable that American social scientists turned to more manageable issues, such as the way in which and extent to which the media alter the traditions and values of newly penetrated countries; the efficacy of the media in achieving political consensus; and the possibility of reinforcing "the impersonal mass media" with the "more face-to-face pattern of social communication."[128] Lucian Pye proposed that American social-science experts in communications were particularly well suited to the task of developing "the appropriate strategies and doctrines which might facilitate the process of nation building in the transitional countries." They had already served as a "remarkably effective bridge between the world of theoretical social science and the world of public policy," concentrating on "the problems of changing the attitudes and behavior of others and of communicating across the barrier of differing cultures" and evidencing "a unique respect for practical policy problems and a high degree of understanding of how scholarship may be turned to policy ends without damaging the growth of knowledge." Born from the wartime need to combat Nazi propaganda, communications research had "continued to be delicately balanced between the scholars' world and that of government officials and private industry," he wrote.[129]

One who achieved such a balance was the political scientist Ithiel de Sola Pool, who, like many other scholars, had for several decades been conducting policy research in the fields of mass communications and United States relations with the socialist and Third World. A 1964 essay by Pool, "The Mass Media and Their Interpersonal Social Function in the Process of Modernization," is representative of his views. Pool had worked under Lasswell's guidance during his graduate-student days and had gone on to concentrate on empirical studies of propaganda, especially Soviet. A longtime professor at MIT, he was a member of the United States Air Force Science Advisory Board and the Defense Department's Science Board. In his essay, Pool said that ruling groups in the developing countries held conflicting views of mass media: in non-Communist countries, they were lukewarm to media development, fearing rising expectations that could not be met; in Communist countries, they used the media to teach discipline and conformity, valuing "the social function of a communication above its truth value."[130] A third approach, with which Pool associated himself, rested on "a criterion of objective truth, which is one of the peculiar features of the Graeco-Roman Western tradition." This approach Pool described as "ruthlessly two-valued and dominated by the law of the excluded middle."[131] He argued that "the Western criterion of truth assumes further that validity can be tested independently of who does the testing."[132] The problem of modernizing was to inculcate this "scientific attitude," to transform values and personality accordingly, and to raise the "achievement motivation."[133]

The kind of media strategies and policy recommendations for the developing world that flowed from social scientists like Pool are well represented in the following summary by Daniel Lerner:

> What is needed in the years ahead is a new conception of public communication as the crucial instrument which can promote psychic mobility [compare Pool's "achievement motivation"], political stability under conditions of societal equilibrium. The mass media can be used to mobilize the energies of living persons (without creating insatiable expectations) by the rational articulation of new interests. Flanked by the schools and community leaders, the mass media can simultaneously induce a new process of socialization among the rising generation that will, among other effects, recruit new participants into political life. The two processes—short-run mobilization and long-run socialization—can then

> converge, a generation later, in the new aggregations or private interests, which are the stuff of a democratic polity.[134]

Such recommendations reveal an advocacy role for mass communication quite different from what media sociologists perceive to be its role in the United States. Furthermore, the sociologists' alignment with the policy maker and their implicit espousal of official government positions depart significantly from the model of scientific objectivity.

By the 1970s, social scientists, especially those who had become specialists in Third World and socialist mass media, began to take up the issues surrounding international broadcasting by space satellite, including the implications that space broadcasting might have for relations between the United States and the Soviet Union and for the accelerating drive by Third World countries to alter the imbalance they perceived in "information flow."

In an analysis of these issues written in 1975, Pool described the history of the controversy over the direct broadcast satellite and took up the criticisms that had been made against the official U.S. position. His earlier assertion about the superiority of the "Graeco-Roman Western tradition" over other cultures is only subliminally evident, although he approvingly restates the typology that had appeared in Lerner's "classic," *The Passing of Traditional Society*.[135] This typology identifies illiterate villagers as "traditional men," migrants to the city as "transitional men," and educated urbanites who "understood the issues of the cold war and ideological movements and followed world affairs with interest" as "modern men."[136]

To begin with, Pool stated that the whole matter was a non-issue, since direct broadcast by satellite "to countries that do not wish it is not an attractive option for either governments or commercial broadcasters."[137] Nevertheless, he examined at length the general problem of "cultural integrity" and its alteration by outside forces, such as "movies, television, literature, songs, hairstyles, clothing styles, patterns of respect and etiquette and religious observances." For Pool, the world was one vast marketplace in which the system that best turns out what the public wants will best succeed in supplying cultural materials. He found that the American system, since it is "predominantly commercial," is the system that "is geared to producing what the public wants." He continued:

> It should be no surprise that in the country where producers try hardest to produce what the public wants, they succeed. The Americanization of world culture so often commented on and deplored might be better described as the discovery of what world cultural tastes actually are, and adoption of those into American media.[138]

Pool cited the U.S. experience in Vietnam in the 1960s as one example of this process. During the Vietnam War, the United States government put into South Vietnam a two-channel television system "to help the American troops feel at home, improve the quality of life and relieve the rural boredom for the population on the government side." One channel was standard U.S. commercial programming; the other was "programmed and produced by the

Vietnamese in Vietnamese and aimed at the rural population." It turned out that in the cities, millions of Vietnamese "turned by preference to the better produced, slicker American programs. . . . Batman won the rating race." Pool concluded that the Vietnamese channel "was not serving the modernized Vietnamese as well as the United States channel." If there were critics of this outcome, Pool warned them not to raise protectionist barriers against foreign materials. Rather, let them take advantage of the situation, since the foreign product from the United States provided "an incentive to producing equivalent, but more locally relevant, material at home." [139]

Pool outlined two possible approaches to solving the dilemma of direct broadcast by satellite, although it was "not a matter of much importance to us." One way would be to yield to the "censorship proposals," partly because the United States was almost alone on the side of freedom, partly because he found (true to his empirical orientation) that "no reality testing [was] available." The other way would be to refuse to cooperate with other governments in "shackling what their peoples can see and hear." It is fairly clear that he preferred the latter course, believing that noncooperation by the United States would make unenforceable any concerted action by other nations: "There is no realistic prospect of the nations of the world compelling us, the satellite launching nation, to enforce a convention we decline to sign." Still, he hoped for a third, or "mixed," approach, by means of which the United States would respond to the anxiety of other nations, avoid censorship, and still "support continued development of satellite communication." Pool's implied threat (of withdrawal of American support from satellite development) was softened by his proposal that the United States "subsidize a flow of communication by satellite from the developing countries to the rest of the world," use U.S. Agency for International Development money to "fund placement of ground stations in poor countries" and "cooperative exchange activities for television programs," back a United Nations "worldwide television network," and provide for extension of the communications power of satellites to include data transmission so that the vast information resources of the West might be accessible to users anywhere in the world. "Overnight, developing countries could close the information gap." [140]

Pool was representative of the postwar, policy-oriented social scientist, and his longtime relationship with the military and intelligence bureaucracy gave his public writings a semiofficial character. The possible contradiction between his claimed scientific neutrality and his unblinking advocacy of United States policies did not appear to enter his thinking. Consider, for example, his designation of the social sciences as "the humanities of our era," a claim he illustrated by citing the effectiveness of social science in "handling man's greatest problem, organized violence. It is the social sciences that give us the most reliable information about where riots will breed, how criminals can be controlled, whether insurgents will maintain rule of a university or a village, and how nuclear war can be avoided." [141]

It is not being argued that the analyses of American social scientists like

Pool are without merit. The problem lies, rather, in their unrecognized assumptions. What may be valid resistance against repression becomes for them "violence" or "riots" or insurgency, and the methods for controlling such forms of resistance are to be found in the humane and rational models of the social sciences. The givens are on the side of the status quo. The very terms—who would be for "violence"?—are selected to win the argument before it starts, to legitimize the policy and the American ideology from which it springs. Other communications theorists describe the same phenomenon (violence) in more subtle language. Daniel Lerner took up the matter in an essay written in honor of Lasswell's life and work as a pacesetter for the management of communication. Lerner identified the media's task as having to cope with two twentieth-century phenomena: first, "*participation,* the rise of public opinion and the spread of media consumption," and second, "*globalism,* the political activation of 'the opinions of mankind.'"[142] Lerner explained: "Coercion cannot forever repress frustration and prevent revolution. Only persuasion—the effective management of communication—can shape public opinion in such manner as to reduce popular frustration and prevent political upheaval. This is the great benefit taught by communication sociology. . . ."[143]

This examination of the study of mass communications in the United States was undertaken in order to discover its usefulness as a guide to understanding the issue of international broadcasting. With due respect to the scholarship and imaginative approaches apparent in the body of work of the major contributors, it must be said that the most revealing information is found between the lines, in the researchers' latent rather than manifest intentions. The early writing on propaganda (before it came to be called mass communications), for example, while purporting to be neutral analysis, actually revealed an underlying anxiety about the possibilities of propaganda as a destabilizing force in the United States. The careful dissection of specific propaganda techniques often proved to be catalogs or guideposts to the achievement of effective word weaponry against unwanted ideas. The propaganda analyst became the counter-propagandist. Later, as more sophisticated research approaches emerged, social scientists appear to have been concerned about the "mass" of mass media. Although the researchers may not have recognized it, we may read in their anxieties the presence of a disaffected population, a segment of society that they viewed as potentially critical of and resistant to the prevailing social structure or at least to its assigned role in that structure.

The eventual partnership between communications research and the military-industrial interests, while causing some small concern among social scientists, was generally accepted as appropriate, despite repeated affirmations about the neutrality of knowledge and scholarship. The unavoidable conclusion is that social science, in general, and the mainstream of sociology of mass communications, in particular, were not neutral, but were shaped by and responsive to their social context. This pattern is still apparent.

Thus as guidance for a study of the emerging world conflict over control of mass broadcasting, the data gathered by and the theories formulated by social scientists in the field of mass communications should be recognized as supporting U.S. government interests in the world arena and the needs and goals of major communication corporations within the United States. This supportive role is little different from the role of social science in socialist countries; but there, the connection between social science and the dominant social institutions is discerned by all, while in the United States, it frequently remains unnoticed.

This insight permits us to understand that U.S. social analysts who examine the controversy over international mass broadcasting and its framework—conflict over news and information flows—largely argue for the official U.S. position without submitting it to the kind of critical analysis that would take them beyond their own conditions.

4

DBS and International Law

The Concept of International Law

The concept of international law, with its ambiguities and uneven history, bears closely on any consideration of international mass communications. Western scholars and students of international law tend to view it as arising historically, out of the needs of states to agree on rules of war, on rules for the mutual use of the high seas, on trade arrangements, and on protecting the rights of nationals. Bilateral treaties were the early form of such law; in the nineteenth century, with the emergence of a more self-conscious nationalism, states with common interests began to make multilateral treaties, as at the Congress of Vienna.

In 1896 at the Hague Conference, the major European nations began to codify international law. A large part of their work dealt with the rules of war, including the clarification of such points as the "opening of hostilities" and the "position in naval war of enemy's merchant ships at beginning of hostilities."[1] Another indication of the increasing importance of international law was the establishment of the Hague Court of Arbitration and of standards for marriage law.

It did not go unnoticed that the emerging body of codified international law was strongly rooted in the interests and needs of the European powers, or, as they were called throughout the nineteenth century, the "Christian" or "civilized" nations. A somewhat less ethnocentric attitude emerged in recognition of Turkey's prowess in the mid-nineteenth century, which won it considerable admiration among European rulers and military strategists, and as a result of Japan's victory over Russia in 1905. In a comment on the possibility that non-Christian states would become participants in the development of international-law traditions, the *Encyclopaedia Britannica* noted that Japan was, fortunately, adopting

> the best she found in Western morals, and in her schools inculcates Christian ethics as a subject *per se* without reference to divine revelation or authority. Turkey too has . . . a code of morals which produces so high a standard of right conduct in private life that very little in the way of moral lessons will have to be learned by the Ottomans from Western civilization.[2]

If it was not easy for the West to make room for others at the diplomatic table, it was even more difficult to explain some of the long-accepted Western rules of war. The notions of the "right of conquest" and that "the victorious are entitled to an indemnity without reference to any question of right and wrong or of justice and injustice"[3] were not easily harmonized with the idea of Christian international law. It could be done, as long as the policies of imperialism and colonial conquest were not widely known. But, observed the legal scholar writing for the *Encyclopaedia Britannica,* "the public opinion of the civilized world," developed by "the national press and periodical literature" and augmented by leader-to-leader and people-to-people contacts, affected "the mental attitudes of nations toward each other."[4] As we have seen, the problem of mass communication, which he was identifying, was to make matters increasingly difficult for the smooth management of international relations.

The double cataclysm of World War I and the Bolshevik Revolution brought about a broad rethinking and reevaluation of the theory and practice of international law. Haltingly, and in spite of the United States' refusal to participate, the League of Nations came into existence and by its very survival, as well as through its multiplying treaties and conventions, hammered out rules for maintaining peace, to be codified beside the longstanding rules for making war. The treaties that had made it internationally "legal" for a colonial power to capture and exploit foreign territories and peoples began to be superseded by at least the idea, if not the practice, of treaties that accorded equal status to all signatories. The publicized claims of the Allied Powers in support of self-determination and the rights of nations to freedom from interference in their internal affairs had helped launch these new principles into the international-law arena.

The horror of World War II again laid bare the destructive potential of a lawless world. In its aftermath, statesmen and jurists of all the victorious powers, this time including the United States, joined to create organizations and write treaties to help regulate international relations without recourse to war. The United Nations, its specialized agencies, and the human-rights principles that oppose racism, colonialism, and genocide and that support world economic cooperation and disarmament were added to the scope and content of international law.

These remarkable developments in the content and direction of international law have been variously assessed by Western jurists. One holds that they are merely disguises: "sublimated power politics."[5] Others have emphasized the accretion of experience in treaty making, peacekeeping, and orderly extension of trade agreements, an accretion that could nurture the "law habit" and perhaps bring about "the long-term advantage of stability and good faith on which international law in the West has largely rested."[6] However, they generally despair of the West's ever finding common ground with the Soviet Union because of the West's deep antagonism against the Soviets' adherence to the principle of collective ownership of the means of production in contrast to the West's fidelity to the principle of private property and the market economy. A third school of Western legal thought, with Philip Jessup among its notable proponents, has argued that nation-states must be more willing to yield up sovereignty to some form of world order: "Until the world achieves some form of international government in which a collective will takes precedence over the individual will of the sovereign state, the ultimate function of law, which is the elimination of force for the solution of human conflicts, will not be fulfilled."[7]

The Soviet jurist G. I. Tunkin argues that the vast changes that have taken place in international law during the twentieth century are to be attributed largely to the influence of the Soviet Union, since its birth in 1917, and to the subsequent national liberation movements and expansion of socialism: "New principles and norms having important significance for ensuring peace and the independence of peoples have emerged; and the old democratic principles and institutes of international law have been developed and strengthened."[8] Tunkin lists the following as new principles: nonaggression, peaceful settlement of disputes, self-determination of peoples, peaceful coexistence, disarmament, respect for human rights, and the prohibition of war propaganda.[9]

In line with his expressed belief in Marx's theory of history, Tunkin regards international law as an aspect of the class struggle, a struggle that will ultimately subvert capitalism and permit the emergence of a new form of social organization. He emphasizes the Soviet interpretation of Marx's theory, which asserts that the class struggle occurs *within* societies and states and is not to be pursued by the use of force by one nation against another. That is, on the international level, the "struggle" is carried out within the framework of "peaceful coexistence": "Struggle does not preclude cooperation," he writes, and "cooperation permeates struggle."[10] It is this that Tunkin regards as the essence of international law. He opposes the Western jurists' espousal of a "world state," deeming it a diversion from the possible, and he condemns the way Western writers downplay the marked change in international law by reducing law to power politics. Emphasizing that he is speaking of the present phase of international-law development, he adds:

> If previously the great imperialist powers frequently declared to be generally recognized norms those norms which they themselves had established, and imposed these norms upon the weaker states, at the present time any attempts of imperialist states to impose norms advantageous to them as norms of international law not only are unlawful but are doomed to failure beforehand.[11]

Both the Western and the Soviet views seem to accept international law as a way of regulating the coexistence of states. The disagreement is deepest precisely over the notion of law as an aspect of class struggle. In the capitalist world, law within a state (sometimes called municipal law) is perceived as a set of binding class-neutral customs or rules adopted by a community of citizens and backed by the force of the state. At the international level, Western jurists see a kind of parallel between internal or municipal, law and international law, which may be viewed as the rules, norms, and customs adopted by a community of nations. The absence of an enforcement agency is regarded as a central peculiarity of international law. This peculiarity is at the heart of the Western ambivalence about the concept. Does it invalidate the idea of international law? reduce it to power politics? Should law be seen as a long process of accrual, only just beginning? For the Soviet legal theorist, accustomed to the historical materialist approach, this peculiarity supports the idea that the formulation and the evolution of contemporary international law are processes of class struggle ("against reactionary bourgeois concepts of international law"),[12] a view that is foreign to and totally rejected by Western jurisprudence.

This brief summary of present-day perceptions of international law underlines not only the profound differences that exist between Western and Soviet jurists, but also the fragility of the concept itself, influenced as it is by the compelling economic, political, and survival contests among twentieth-century states.

International Law in Outer Space

Despite their different philosophical and juridical outlooks, socialist states and capitalist states have been making treaties and having diplomatic relations for many decades.[13] Advancing technology, including the penetration of outer space, has been a major factor in widening the arena in which state interests meet and must be accommodated if open conflict is to be avoided. The destructive nature of nuclear war, a "space activity," has profoundly affected the interest of peoples and their leaders in negotiating peaceful solutions to their disagreements. In addition, the technical and scientific knowledge that has spawned electronic communication on a world scale, another "space activity," has made it possible for masses of people to become more aware of their condition and more informed of the efforts and actions of their leaders. These and other factors have led to efforts to bring these new arenas of human activity under the aegis of international law.

In the early years of discussion about space law, there were few allusions to the specific problem of DBS. The focus was on the legal status of outer space. Does existing international law apply in outer space? Both Soviet and Western scholars agreed that it does, although their supporting arguments showed quite different degrees of enthusiasm. The Soviet view from the beginning has been that the extension of international law into outer space is fundamental. Writing in 1959, the Soviet jurist E. Korovin[14] argued that while

> national sovereignty does not extend to cosmic space, it does not follow that space is to be considered as some kind of legal "vacuum" where no restrictions on freedom of action prevail. All universally accepted rules of international law (inadmissibility of the use of force in solving international disputes, non-injury of foreign citizens and their property, governmental responsiblity for the activities of their representatives, etc.) apply to the Cosmos as well.[15]

We have commented on the ambiguity that characterizes perceptions of the meaning of international law. If it embodies, as the Soviet experts assert, such "new principles and norms" as nonaggression, peaceful settlement of disputes, self-determination of peoples, peaceful coexistence, disarmament, respect for human rights, and the prohibition of war propaganda, then the extension of such principles to cosmic space would provide a framework for the regulation of DBS as well as of all other space activities undertaken or to be undertaken by states. But Western jurists are less certain and specific in their commentaries. After an exhaustive study of the literature of space law, C. Wilfred Jenks concluded that it contains, perhaps "inevitably," an excess of caution and of fancy, although he felt that these qualities are balanced by practicality. On the question of whether the jurisdiction of international law extends into outer space, he himself tended to evidence an excess of caution. Rather than examining the question directly, he buried it within an analysis of the principle that space is free for exploration and use by all states and is not subject to national appropriation. This principle, he felt, had been a general theme in space-law literature and by 1965 had been "placed beyond the pale of reasonable controversy" as a result of having received the General Assembly's unaminous endorsement in 1961 and 1962. After considerable discussion of related matters, he again turned his attention to the issue of international law being extended into outer space. His comments point up the uncertainties and unresolved nature of contemporary international law, reflecting as it does the continuing political, military, and class tensions of our time.

> It is perhaps sufficient to say that it is unnecessary to project into space the innumerable doctrinal controversies of the theory of international law and that our objective should be rather to thread our way among the difficulties presented by differences of a doctrinal nature with a view to assisting in preparing the way for the practical solution of concrete problems.[16]

Possibly having in mind the definitions of international law repeatedly given by Soviet jurists, Jenks went on:

> One of the questions which [remains] unresolved may be of more than doctrinal importance, namely the question of how much firm law can already be said to exist. But as the effect of attempting to define how well-established the law has become is frequently to weaken rather than to strengthen the degree in which a sense of legal obligation already exists it would seem preferable to leave the determination of the measure in which particular principles or rules have become authoritative in character to be settled as need arises by the practice of states and future treaties and judicious decisions.[17]

The practice of the International Telecommunication Union, which establishes international norms that have the authoritative character of the laws of states, may be an illustration of Jenks's preference. Its history as an international "lawmaking" body is known to us. Although it began its work long before the emergence of space law, its more recent actions can be viewed as having extended the reach of international law into outer space on a de facto basis. For example, in 1959, the ITU allocated radio-frequency bands for remote-control space vehicles (Bloomfield called this "the first concrete regulatory measure in the outer space field"),[18] and it has continued to broaden its jurisdiction over frequencies used for space communications, including broadcasting.[19] True, ITU regulations, like most of international law, are not enforceable. States are not obliged to adopt them, and adherence to the ITU's prescribed spectrum allocations is voluntary.

In a series of lectures given in 1968 on international law and the uses of outer space, the prominent British jurist J. E. S. Fawcett commented on the jurisdiction of the ITU over satellite communication, although he did not specifically cite the ITU jurisdiction as an extension of international law to outer space. Possibly, he preferred a functional approach—that is, the accumulating of law designed to regulate a particular function, in this case international uses of radio—thus avoiding the issue of a general extension of international law.

The accumulation of a body of regulations dealing with an international activity like radio communication is noted by some jurists as an example of how the idea and substance of international law may be strengthened, thus diminishing the likelihood of conflict. Oliver Lissitzyn warily suggests that joint activities in outer space will further the necessary "reinterpretation of Communist ideology" and will slowly erode the Soviets' "extreme hostility to other systems of public order,"[20] which stands "in the way of the gradual evolution of mankind toward the rule of law in world affairs."[21] Edward McWhinney, a Canadian authority, sees strong potential for a cooperative approach "between states with opposing social systems" if they hold to "modest and low level, problem-oriented methods of resolution of current international tension-issues."[22]

Georg Schwarzenberger, on the contrary, believes that functionalism is doomed to fail because "the governments and bureaucracies of the World Powers are little inclined to share their governmental and administrative monopolies with other government representatives and competing sets of international bureaucrats." He disparages the specialized agencies of the United Nations as examples of just how far states are willing to go: "[They] confederate international organizations with essentially consultative and prelegislative functions and, within a margin of tolerance, let them behave as if they were the organs of a functional international community."[23]

Philip Jessup also sees "inherent limitation" to functionalism; although it can be very productive, it cannot substitute for political solutions.[24] Building on what he perceived as "the United Nations habit,"[25] Jessup proposed an international regime for space, with responsibility for its administration to be vested in the United Nations General Assembly or in a new international body with legislative power.

In 1963, the United Nations General Assembly unanimously adopted the Declaration of Legal Principles Governing the Activities of States in the Exploration and Use of Outer Space, a first draft of which had been submitted by the Soviet Union in the 1962 session.[26] Significant provisions included the principle that international law, including the United Nations Charter, applies to outer space and celestial bodies and that outer space and celestial bodies are free for exploration and use by all states in conformity with international law and are not subject to national appropriation.[27] After several years of disagreement, the United States and the Soviet Union joined the other members of the General Assembly in creating the Committee on Peaceful Uses of Outer Space (COPUOS), which was to be made up of twenty-eight member states, including the United States and the Soviet Union. The General Assembly designated one of the two subcommittees of COPUOS as the world forum for consideration of legal problems arising from the penetration of outer space, a task that it has continued to perform. By 1967, these undertakings had led to a breakthrough international agreement—the Treaty on Principles Governing the Activities of States in the Exploration and Use of Outer Space, Including the Moon and Other Celestial Bodies (or the Outer Space Treaty)—which was signed by all the major powers simultaneously in Moscow, London, and Washington.

In general, the evolution of international law regarding space has paid attention to the following problems:

1. How to establish legal jurisdiction over something that, by treaty agreement, is not to be appropriated by states or individuals (corporations).
2. How to legally reconcile the concept of private economic interest and ownership with the concept of nonownership or collective ownership.[28]
3. How to protect the national interest of states from hostile acts of other states. Put another way, how to use outer space for the preservation of peace and to prevent its use for war.

State Sovereignty in Space

Coupled with the matter of the applicability of international law to outer space is a second large issue: the question of state sovereignty in space. The Outer Space Treaty provides for free use of outer space in accordance with international law, prohibits national appropriation of outer space, and makes states the sole responsible entities for observing and enforcing its provisions.

In examining the meaning of this complex duality—freedom to use but no freedom to appropriate—it is useful to review the struggle over control of air space, which goes back to the Paris air conference of 1919 and the years of the League of Nations. During the league's efforts at disarmament in the 1920s, there arose support for "world air transport under an international authority" as one way of accomplishing "air disarmament";[29] that is, of preventing the possibility of foreign aircraft entering the air space of unwilling countries. It is an interesting sidelight, foreshadowing the views of some present-day space lawyers who argue in favor of a global system of international communications as a means of avoiding the problem of sovereignty with regard to DBS.

Another foreshadowing occurred at the convention of the International Civil Aeronautics Organization (ICAO), held in Chicago in 1944. The conferees took up the question of state sovereignty in the air space above the territory of a state. The United States sought unsuccessfully to establish the right of commercial airplanes to freely use air space without regard to state sovereignty:

> At the conference in Chicago that preceded the adoption of the Chicago Convention of 1944, the United States had pressed initially for a central principle of the freedom of the air, to be concretized in one overarching body of international law principles and to be institutionalized in the one paramount international agency.[30]

If such a policy had been adopted, a world air-transport system might have evolved under the ownership and control of the few Western European states and United States corporations then in a position to move forward. But the U.S. position did not prevail in Chicago; the ICAO agreement reconfirmed the sovereignty of a state over the air space above its territory (which had been accepted originally in 1919). One of the results has been the proliferation of separate air-transport systems, most of which are state owned (except in the United States, where they are state subsidized but privately owned).[31]

The principle of state sovereignty over air space has been explained by John C. Cooper, an international lawyer and, from 1934 to 1945, a vice president of Pan American Airlines:

> For many years, the legal principle has been accepted that each state is sovereign in the airspace over its national lands and waters, that its unilaterally determined laws and regulations apply there to the exclusion of all others and that it has the sole and exclusive right to permit, control, or deny flight in its airspace.[32]

Cooper also wrote "that the territory of a State shall be understood as including the national territory, both that of the mother country and of the colonies, and the territorial waters adjacent thereto."[33]

The development of space satellites forced a reexamination of the principle of state sovereignty over air space and of its implications for the legal status of outer space. An immediate issue was where air space ended and outer space began. Cooper argued in many forums that existing air law should be clearly distinguished from emerging space law by setting "an upper limit to national sovereignty, with its rights of unilateral regulation."[34] If this question were resolved, states could then agree on the legal status of outer space, which, Cooper urged, should be treated similarly to that of the high seas.[35] Cooper's recommendation to establish an official cutoff point between air space and outer space was much discussed but not adopted. In its place evolved a kind of de facto agreement, which sets the altitude of airspace sovereignty at that level above which aerodynamic flight is not feasible and below which orbital flight is not feasible.

In 1959, the Soviet jurist Korovin took up the question of sovereignty in space, arguing the inapplicability of not only Roman civil law ("whose the land, his also the sky") and air law (from the Paris and Chicago treaties), but also the law of private property, laws of individual countries, maritime law, and the law of the Antarctic. He also rejected the interpretation of some Western jurists that "since no country has protested against the launching of Soviet and U.S. earth satellites this may be interpreted as 'freedom of outer space' established by universally recognized precedent." Korovin argued that since the Soviet Union ("just as the USA") had put its satellites into outer space under the programs of the International Geophysical Year in 1957, there was, in effect, "preliminary consent of all the countries concerned."[36] If the concept of sovereignty was not applicable in outer space, as Korovin insisted, there were nonetheless rules that he felt did apply, and these rules derived from existing treaties and agreements made by states and enforced by them: "The presence or absence of separate specialized legal systems . . . cannot abolish the generally recognized principles of law prevailing in our time wherever people are active on land, sea, air or space." He continued, "In this, and only in this sense, can one speak of 'freedom of space' or 'open space', that is, the right of each country to use cosmic space as it sees fit without doing harm or causing injury to other states."[37]

Other Soviet commentators have elaborated on Korovin's early comments, rejecting the implication that outer space is somehow beyond sovereignty and insisting that "the basic principles and rules of international law extend to space activities too."[38] A. S. Piradov, editor of a Soviet book entitled *International Space Law,* asserted that activities in outer space, like all other human activities, should be guided "by the interests of humanity rather than the specific physical properties of this or that sphere, or the peculiarities of the technological means employed to master the sphere." The "interests of humanity," Piradov argued, include the contemporary international-law principles of "equality, peaceful coexistence, and cooperation of states, non-interference in one another's internal affairs, etc."[39]

A Western view of the problem of sovereignty in space was set forth in 1961 by the political scientist Lincoln Bloomfield at a conference on outer space held by the American Assembly. He reviewed various international-law definitions of sovereignty, starting with the Papal Bull of 1493, which sought to divide the New World between Spain and Portugal, and including the "hinterland" principle, which in the sixteenth and seventeenth centuries legitimized the practice of colonizers in America and Africa by asserting that those who claimed the coast were automatically assigned sovereignty rights over the contiguous inland regions. Tactfully silent on the legal justifications for the nineteenth-century wars of conquest and empire building—including the agreement made at the Berlin Conference of 1885, by which European states simply awarded themselves sovereignty over most of Africa—Bloomfield completed his historical review with a reference to the twentieth-century evolution of "the doctrine of effective occupation," under which "sovereignty requires possession and also occupation with intent to make it permanent."[40]

None of these once-acceptable international laws seemed to Bloomfield to be applicable to outer space, and he turned to the suggestions of Western jurists that space "not be considered susceptible to appropriation by any nation."[41] Noting that presidents Eisenhower and Kennedy and United Nations Secretary General Dag Hammerskjöld had proposed that there be no sovereign claims in space, Bloomfield recommended that this idea be incorporated into a treaty and regretted the tendency of Western statesmen "to leave these matters to chance." He expressed his belief that such a principle was becoming established de facto as a result of the satellites already in orbit: "Ever since Oct. 4, 1957 . . . a precedent has existed for the free use of outer space without objection by the state over which the satellite has passed."[42]

As we know, the extension of state sovereignty into outer space appears to have been prohibited by the 1967 Outer Space Treaty, but a number of legal questions about the matter, particularly in the area of broadcasting, continue to be examined by international lawyers. In one of the many articles that were published in Western, primarily United States, law journals during the first decades of the space age, M. S. Geston raised the question of how a nation or bloc could be prevented from transmitting broadcasts indiscriminately. Indirect action by the offended state could include trade sanctions and diplomatic protests; direct action could include jamming.[43] Geston felt that the legal justification for such action might be based on "a theory of penetration by the radio signal of the sovereign territory," a view that does not appear to touch directly on the question of sovereignty in outer space. However, Geston thought that defensive jamming might be augmented by going after the satellite itself, either by interfering with the satellite-to-earth signals or by impairing the satellite. When Geston was writing, in 1970, the drastic implications of this course were not a "realistic alternative," since only two or three nations had the capability to accomplish it. But he also added that the other alternative, jamming of the incoming signals from space, was relatively ineffective. (Geston explained that jamming works against shortwave radio and is used by a number of states to block out unwanted broadcasts, but it is extremely difficult to jam signals beamed from a satellite; to jam such signals is like trying to flash a flashlight directly into the reflector of another flashlight.)[44]

A searching examination of the issue of sovereignty in international space communication was made in 1971 by Charles Henry Alexandrowicz, who found sovereign-mindedness to be "the determining element in the definition of the legal status of the space in which global communication services operate."[45] Alexandrowicz cited precedents set by the international law that governs civil aviation, noting that the ICAO agreement of 1944 had provided that only states were to have the power to register aircraft. From this, it followed that the laws of the registering state governed events occurring inside the airborne plane, as long as these events produced "no effects outside the vehicle."[46] As lawlessness (presumably highjacking) on aircraft became a concern, provisions were added in 1964 to the ICAO agreement, permitting the application of the laws of the territory over which an unlawful act occurred. "In this respect," Alexandrowicz noted,

> the situation is different from the legal regime on board ship, which follows the law of the flag and presumably on board satellite, which, apart from the short period of launching and splashdown, finds itself in outer space beyond the reach of state sovereignty and thus follows the law of registration.[47]

Alexandrowicz asked why global communication "should be treated separately from regional or inland communication"; in other words, why the issue of state sovereignty over space communications should be subject to international rather than national (internal) law: "The answer is that world communication is a social reality which, because of its indivisibility, is of special significance involving the whole family of nations and the law of nations as a universal legal system, irrespective of geography or political ideology."[48] The regulations and jurisdiction of the ITU are presumably seen as part of that "law of nations":

> It can be inferred from the advisory opinion of the International Court of Justice . . . that if the vast majority of the family of nations adheres to an organization the latter enjoys the benefits of objective international personality. Its personality in international law is valid not only vis-à-vis members but *erga omnes,* with global legal validity. This rule applies not only to the United Nations but also to the specialized agencies.[49]

Alexandrowicz was aware of the tenuous "paralegal" character of many of the regulations of the specialized agencies, including the ITU. Lacking enforcement power, ITU requirements are not legally binding, although there is always the possibility of expelling the offending state from the United Nations (membership in the ITU is conditional on membership in the United Nations) or from the ITU (at the centenary ITU meeting in Montreux in 1965, member states voted to expell the Republic of South Africa). Alexandrowicz suggested that this evolving paralegality "may indicate a wider trend in the development of contemporary international law, that is, the growing and often inescapable dependence of states on participation in activities of global and indivisible concern, where the principal sanction is exclusion from participation of those states that refuse to comply with universally accepted standards."[50]

In sum, there appears to be consensus that international law precludes the making of territorial claims in outer space. In space vehicles, including communications satellites, the laws of the state of registration apply. But when spacecraft, including electronic signals that emanate from them, affect states other than the state of registration, international law applies. Presumably, this international law includes the regulations of the ITU and the provisions of the United Nations Charter, the Outer Space Treaty, and other international treaties that are applicable. While the Soviet theorists assert this as a fundamental principle, the Western authorities appear to regard it as arising out of actual conditions.

Individual Autonomy in Outer Space

Closely related to the issue of state sovereignty in space is the question of individual actors and the degree to which they are autonomous when they are in outer space. The pertinence of the question for DBS arises out of the private nature and presumed autonomy of enterprise in some capitalist states; jurists have repeatedly taken note of the possibility of private communications corporations conducting broadcasts via space satellites, and the history of shortwave international broadcasting records a small number of U.S.-based private ventures, some religious and some conducted for private profit.

From the beginning, Soviet jurists have held that states are the only entities that are subject to international law. Alluding to the problem of damages that might be caused by space objects, Korovin asserted in 1959 that since launching of satellites was "under present conditions . . . solely under the auspices of governmental bodies, full responsibility for eventual damage lies with the government concerned."[51] This assumed either that private bodies would not get into the business of launching or operating satellites or that, if they did, the state in which they were located would assume responsibility for them. In 1963, this view became one of the principles adopted by the United Nations, and in 1967 it became Article VI of the Outer Space Treaty.

The Hungarian space-law authority Gyula Gal expanded on this issue in 1969.[52] Noting Article VI of the treaty, he stated that "insofar as . . . international law is concerned, the subjects of space law can be only states and international organizations." Further, "international space law will never directly confer rights or obligations on citizens."[53] Gal provided a brief history of the framing of the principle that deals with space activities by private bodies. The Soviet draft had provided that only states could explore and use outer space. The British draft would have allowed equal exploratory and use rights to all states and their citizens. During the debate, there were allegations that the Soviet proposal was designed to undermine the economic system of other states: "Canadian delegate Cadieux . . . explained that it would be impossible for many states to deprive a private enterprise of the expected—considerable—commercial profit of the communications system, without prejudicing their own principles."[54] Gal concluded: "When the legal system of a state permits space activities on the international plane by private organizations, such activities must be carried on in accordance with the rules of international law and on the responsibility of the state capable of granting authorization and obliged to exercise supervision."[55]

Gal's description of the problem iterates the policies of the Soviet Union. It is the kind of analysis that is rare in Western writing on the subject:

> Freedom of use and exploration of outer space is vested in states. The Space Treaty leaves no doubt that only states can be the subjects of this freedom—as a right of international law. Neither the Treaty nor the customary-law rules

> confirmed by it can directly empower individuals or private organizations. In one of the states carrying on space activities—in the Soviet Union—the social-economic system itself excludes the possibility of private organizations participating in space ventures. Not so is the case of other space powers. For example, in the United States, in accordance with the economic system prevailing there, private companies play an ever increasing role. Thus, for instance, the Telstar communications satellite launched July 10th, 1963, was the joint venture of NASA and ATT (American Telephone and Telegraph Co). Not long thereafter the U.S. legislature passed President Kennedy's bill which was enacted as the Communications Satellite Acts of 31 August, 1962. The Act enables any such commercial company to operate a cosmic telecommunications network as can avail itself of the cooperation with a state authority—the NASA—for the launching of the satellites.[56]

We are already familiar with this issue, beginning with the St. Petersburg conference in 1875, when the United States did not sign the International Telegraph Convention because to do so would have had the effect of assuming state regulatory power over the private telegraph company or companies that operated in the United States. The same stumbling block kept reappearing at each new treaty-making conference and had profound effects on the various telegraphy and radio agreements. But by 1967, when the United States signed the Outer Space Treaty, world power arrangements, the needs of the space industry, and international public opinion were important factors in mooting the issue. For those reasons and perhaps for others, the United States became a party to the Outer Space Treaty and thereby assumed regulatory responsibility over private entities in the United States that were engaged in space industries, including not only space communications, but also the emerging fields of scientific experimentation, commercial production, and projected mining and colonizing activities in space or on the moon.

Many believe that the ascendancy of more conservative leaders in capitalist countries during the past decade has brought a marked change in the willingness of the United States to participate in international lawmaking arrangements that are not fully under its own control, that appear to affect its commitment to private enterprise, or that imply a diminution of state sovereignty in favor of collective or international action. One example is the so-called Mining of the Moon Treaty, which the United States and other major powers negotiated for a number of years but which was eventually defeated by corporate interests that opposed it as an obstacle to long-range planning for profitable mining operations on the moon. Another example is the Law of the Sea, which, during ten years of negotiation, proceeded in the direction of prohibiting sovereign claims to the seabed and creating instead a United Nations–related agency to control the mining of the seabeds and to distribute the gains among member states, including the underdeveloped and the landlocked. Less than two months after assuming office, the Reagan administration withdrew United States' approval of the progress that had been made in these negotiations. "No national interest could justify [yielding] sovereign control of two-thirds of the earth's surface to the Third World," commented Leigh S. Ratiner, a member of the American negotiation team.[57] In August 1982, the treaty was adopted, without the approval of the United States. Had President Reagan and others like him been speaking for corporate capitalism in 1967, it seems doubtful that the United States would have been a signatory to the Outer Space Treaty, which was not then seen in such a confrontational light.

DBS as a Subject of International Law

The foregoing review of some of the main legal issues arising out of space activities provides background for consideration of the controversy over direct broadcasting by satellite. The question over whether and how this technology was to be internationally controlled came to a head in 1972, when by a vote of 101 to 1, the United Nations General Assembly asked COPUOS to "elaborate principles governing the use by states of artificial earth satellites for direct television broadcasting with a view to concluding an international agreement or agreements."[58]

We have reviewed the rise of broadcasting across national borders and its international regulation by treaty, including the conventions of 1906 and 1912 and the decision in 1932 to combine the International Radiotelegraph Union with the International Telegraph Union to form the International Telecommunication Union.

Except for the League of Nations treaty in 1936, which sought to limit the use of radio propaganda by one state against another, there has been no international effort to establish or limit state sovereignty over foreign radio signals that penetrate the air space of a state. Nor has there been discussion of any principle that would prohibit national appropriation of radio air space by states. The problem of interference among radio signals has been approached by international agreements to assign to the ITU the technical tasks of designating spectrum segments for certain uses, setting standards, and maintaining a public register of all broadcasting frequencies in use and their users. The unspoken assumption of this ITU jurisdiction, which has existed for a relatively long time and has thereby possibly gained the status of customary law, is that the space through which radio waves travel is free for the use of any state, restricted only by the technical requirements of the ITU.

The possibility of broadcasting via space satellite caused the issue to be viewed very differently, both from a technical standpoint and in the larger context of world history. In the years since the end of World War II, new conditions had arisen that were having profound effects on the politics and technology of telecommunication. Foremost among them was the undreamed-of rate of development of space-communications services. Of almost greater significance was the dominant role—"leader of the free world"—that had been assumed by the United States, in space science and communication, as well as in military power, political influence, and commercial expansion. A third new condition of greater importance was the emergence of new forces in the world political arena—the countries of the Third World and the noncapitalist countries. The United Nations and its specialized agencies provided them with a forum for achieving their goals, which included a greater voice in managing and benefitting from the world's resources.

In addition to these factors, there were technical reasons for identifying DBS as being different from terrestrially transmitted radio. The communication satellite has the capability of transmitting television signals as well as radio, and television is generally perceived as a much more powerful means of communication and persuasion than radio. Further, communication satellites permit a far greater "reach" than radio, thus eliminating the politically difficult and financially costly radio-relay ground stations in foreign countries, which are now used by the international broadcasting systems of the United States, the United Kingdom, the Soviet Union, and other states. Also, since they are in the microwave range, signals transmitted via satellite are relatively free from the natural interference that distorts radio shortwave signals. Finally, as we have noted, satellite-to-earth signals are relatively invulnerable to jamming.

The advantages of satellite broadcasting as contrasted with terrestrial broadcasting caused states to consider the matter of international regulations anew. For poorer countries (the majority of the world), the idea of direct broadcast by satellite seemed full of promise, as a way to combat illiteracy, improve public health, promote agriculture and industry, and build national consciousness. However, the launching of a satellite requires expensive rocket technology, which has been developed by only a handful of nations. Without some form of international regulations, many countries saw themselves as cut off from the valuable potential of DBS and at the same time extremely vulnerable to satellite penetration by external interests that would not necessarily be consonant with their own. Many nations were concerned specifically to establish some form of prior-consent arrangement before television signals from another country could be beamed by satellite directly to their populace.

As we have seen, satellite-transmitted communication signals, whether television or radio, were from the beginning assumed to come under ITU jurisdiction, presumably because of the long ITU history of regulatory authority over international telecommunication. But what of the fact that such signals move through outer space? Did that make them subject not only to the ITU, but also to international space law in general? J. E. S. Fawcett thought that it did: "Commercial broadcasting by satellite would be an activity in outer space" and thus would be covered by the Outer Space Treaty, including the provision that a state that had signed the treaty is accountable for broadcasting enterprises conducted by its nationals.[59] S. Houston Lay and Howard Taubenfeld appear to agree: "The usual rules relating to electromagnetic communications apply to the new technology, and the International Telecommunication Union has allocated frequencies for space communications in accordance with its customary procedures."[60]

As the study and use of space satellites communications have proceeded,

Fawcett's interpretation has not been widely contested, although the continuing political disagreement over the regulation of DBS makes it evident that the issue is still unresolved. The question is: Since broadcasts via satellite involve space, should such broadcasting activity be subject to international space law (the Outer Space Treaty) as well as to the international terrestrial law of the ITU? The difference may be more significant than it appears.

In his study of the legal status of communication via space satellite, Alexandrowicz asserted that "the concept of *res communis* is now a *fait accompli* in relation to radio space."[61] He argued that radio space telecommunications have a legal status similar to that of traditional radio, including shortwave. This is "felicitous," he felt, since it means that "legal development has made it possible to fit outer space activities into the institutionalized legal framework of the spectrum mechanism."[62] Alexandrowicz speculated that radio space might have evolved differently had it "been legally assimilated to air space." While novel, it is possible to construct an argument for this possibility: if a state exercises sovereignty over its territory, waterways, and air space, why not over its sound waves? Alexandrowicz sought to undermine any such claim by emphasizing what he regarded as the conditions of radio communication through space that had prevailed since the launching of *Sputnik:* "While the task of defining the legal status of radio space has been fraught with considerable difficulties, which have taken much time to overcome . . . , the definition of the legal status of outer space was consented to by the powers as soon as satellite communication became a reality."[63]

Alexandrowicz went on to compare the more advantaged status of outer-space radio communication with the very different status of air space, which, as we have seen, is, by international law (the ICAO treaty), subject to state sovereignty. He argued that national control of air space has been a stumbling block to multilateralism and to improved distribution of world services. With its sovereignty over its air space protected by international law, a state is able to bargain with it in exchange for "reciprocal concessions."[64] Alexandrowicz was referring, one imagines, to the needs of the market economy, which he saw as transcending assertions of sovereignty if such assertions are obstacles to expansion. Carl Christol, in an analysis prepared for the Naval War College series on international law, took a view similar to that of Alexandrowicz.[65] He considered the distinction between radio space and air space in the context of a state's right to jam an unwanted signal. It certainly cannot do so on the basis of its established sovereignty over air space, since air-space boundaries do not control radio waves. (But if it were a question of national security, there might be justification for jamming—based on another kind of sovereignty, the inherent right of self-defense.)[66]

As the members of the United Nations grew more concerned about the implications of DBS, a concern that accelerated in proportion to the rapid development of satellite-communication capability in the United States, the discussion of the status of DBS in international space law became more specific. At the 1972 IISL colloquium, Soviet jurist Yuri Kolossov took up the "seeming contradiction between the principle of the sovereign right of the state to regulate its own broadcast services and the principle of non-interference against political and cultural elements of other states."[67] He felt that there is no contradiction: noninterference in other states' affairs is "lex generalis," and state sovereignty over internal broadcast services is "lex specialis," the former apparently being of a higher order and thus legalizing the action of a state to counteract unwanted foreign broadcasts. He enumerated possible counteractive measures: intergovernmental consultations; international agreement prohibiting the broadcast of material containing "ideas of war, militarism, racial hatred, apartheid, sadism, pornography and similar ideas condemned by modern international law"; and no intentional satellite broadcast from foreign territories "without prior clearly expressed consent . . . on the part of the government exercising jurisdiction over the corresponding territory." Kolossov argued that Article IX of the Outer Space Treaty "is applicable to the telecommunications via satellite" because it obliges the signatories "to conduct all their activities in outer space with due regard to the corresponding interests of all other" signatories, a provision that, he felt, mandates international consultations.[68]

In summary, on the question of whether electromagnetic signals transmitted across national borders via space satellite are subject to controls that are different from those applicable to such signals when transmitted terrestrially, we see a pattern of legal views not unlike that already established during the arguments over the extension of international law to space issues, the problem of state sovereignty in space, and the legitimacy of private entrepreneurs in space. Western authorities tend to downplay the question and to promote a free-trade view comparable with that underlying shortwave transmissions. Soviets underline the applicability of the broad range of emerging international legal measures, including the Outer Space Treaty and the United Nations Charter. The post–World War II history of international broadcasting via shortwave, including the ongoing effort of the United States government's Radio Free Europe and Radio Liberty to undermine the socialist states, is no doubt reflected in the Soviet determination to put DBS in a different category from shortwave.

The Military Question

Overarching the international legal status of direct broadcast by satellite is the little-known dimension of its military uses. We have seen that international law has evolved over centuries partly as rule making for warfare and more recently as rule making to avert war, the latter purpose now dominating the pages of international-law literature. Legal theorists understandably prefer to dwell on their achievements as peacekeepers and to try to build a legal framework that will preclude nuclear war and the end of livable existence on earth. That may be why the space-communications literature so rarely mentions the vast military uses of satellite communications.

The member states of the ITU exempt military applications of electromagnetic broadcasting techniques from their registration requirements and from any other controls they might agree on. J. E. S. Fawcett is one of the few legal authorities who have pointed out that the ITU makes no conditions or restrictions on military radio installations of its members.[69] Not that the literature is oblivious to the military uses of satellites. M. S. McDougal, Harold D. Lasswell, and I. A. Vlasic itemized them in their book *Law and Public Order in Space:* reconnaissance, geodetic and meteorological observation, communications, navigation, missile warning, and mapping.[70] While the writers did not mention the ITU exclusion, they did give special attention to the communications satellite, noting its superiority over conventional modes of communication and concluding that a satellite communications system "obviously augments the bases of power of a state."[71]

The requirement in the Outer Space Treaty that all activities in outer space be "peaceful" touches on this question of restrictions on the military use of space satellites. Few commentaries on space law discuss the treaty in this connection. Two authorities who did, in 1970, cited the divergence in interpretation of the word *peaceful:*

> Both the Soviet Union and the United States are in agreement that outer space should be used exclusively for peaceful purposes. Both have thus far classified all of their own missions as "peaceful." Here, a divergence in language occurs if, perhaps, not in practice. The Soviet Union, for its definition of peaceful mission, rules out all "military" activities. It has thus charged that United States high-altitude nuclear tests, observation satellites, and the like, are military, aggressive, and contrary to the principle of peaceful use, although the Soviet Union has itself done all these things as well. The United States, on the other hand, argues that "peaceful" means only nonaggressive and not necessarily nonmilitary. Hence, military activities which are, on one basis or another, not hostile or aggressive, are permissible.[72]

The continued existence of states and their overriding interest in survival make the formulation and discussion of international law, whether space law or terrestrial law, an uncertain activity. We would do well to recognize this as our study proceeds and thus avoid the prevalent practice of subdividing the world into a civilian portion, to be studied and managed, and a military portion, to be left unscrutinized and uncontrolled. Unfortunately, mainstream social analysts have not given the matter of the militarization of space much consideration, although with the rising intensity of the arms race, including the U.S. government's Strategic Defense Initiative ("Star Wars"), more studies are now becoming available.[73]

A Single Worldwide System

A recurring theme in space-law literature has been the possibility of tying the whole world into a single communications system, an idea that recalls Woodrow Wilson's call for a worldwide radio-communication system. The post-*Sputnik* years saw a revival of the concept in some of the documents generated by the United Nations—for example, the preamble to an early resolution on outer space, which states that communications by satellite should be available on a global and nondiscriminatory basis. Significantly, the manner of control of such a system was not touched on, except that it was to be achieved through international cooperation.

Eisenhower, in 1960, spoke at the United Nations in favor of a worldwide communication service, as did Kennedy, in 1962. Khrushchev also proposed "examining the possibilities of cooperation" in establishing worldwide communications systems.[74] And as the issue of unwanted DBS signals began to take shape during the course of ongoing discussion and debate, the possibility of one "worldwide" system gained popularity. It seemed to provide a way out, although the specifics of such a system varied greatly, depending on its proponents.

As already noted, in the discussion of international law, the idea of ending conflict by establishing a world government seems to have attracted more commentary from the West than from the Soviet Union or other socialist countries, which have repeatedly emphasized the importance of states as the irreducible fundamental units in international law. Perhaps the difference is due to the fact that, at least until the mid-1970s, the West seemed to be in a position of dominance for whatever negotiations might precede the formulation of steps toward a multinational or supranational approach. The West could afford to theorize about globalism, since it very likely would be able to control the outcome of such considerations. Authorities from the Soviet Union, viewing themselves as encircled or even embattled, have tended to disregard or reject outright such theorizing as utopian at best and as calculated Western expansionism at worst. Not that there is unanimity among Western jurists on the question of international world order: views range from the extreme concept of one world without national sovereignty, to the more moderate call for "an integrated, contextual, and future-oriented approach," to the incremental approach, "the patient work of statesmen" solving concrete problems by developing proposals to present to states for their acceptance.[75] For some, discussion of a world system took a very practical turn as an argument in favor of a planned initiative to create such a system under the control of the United States, as was done in 1962 when Congress created the Communications Satellite Corporation (COMSAT). As Edwin J. Istvan, director of international development for COMSAT, put it in 1966: "Technically, a single worldwide system provides the most effective use and management of the limited frequency spectrum; avoids duplication of and interference between competing systems; improves operating efficiency; and reduces the technical and operating problems of compatibility between different space systems and other services."[76] The founding legislation describes the rationale for and the special nature of the corporation that Congress had created:

> The purpose of the Comsat Act was to establish, in conjunction and in cooperation with other countries as expeditiously as practicable a commercial communications network, which will be responsive to public needs and national

> objectives, which will serve the communication needs of the United States and other countries, and which will contribute to world peace and understanding. In order to facilitate this development and to provide for the widest possible participation by private enterprise, United States participation in the global system shall be in the form of a private corporation, subject to appropriate governmental regulation.[77]

COMSAT moved expeditiously in furtherance of the goal of a global system. In 1964, it instituted the International Telecommunications Satellite Consortium (Intelsat), inviting other states to buy shares in the new enterprise. By 1968, some 58 nations had become members, and by the 1980s, there were 109 members. The preamble to the Intelsat agreement expresses the signatories' desire to set up a "single global commercial communications satellite system . . . which will contribute to world peace and understanding."[78] There were difficulties, however. Bloomfield had foreseen some of them in 1962 when he commented on the "eventual development of a well-integrated global system of communication satellites linking the world by telephone, telegraph, radio and television." He pointed to the problem of how other countries might feel about the assignment of space radio frequencies to meet U.S. needs at the expense of others, and to the political "quid pro quo" that is implied by United States participation in "an international agency in which other countries have an appropriate voice."[79]

By 1967, when Hungarian jurist Gal wrote his study of space law, COMSAT and Intelsat were well advanced, with at least five communications satellites stationed in orbit and providing communication connections across the Pacific and Atlantic oceans. Gal felt that the satellites were leading not toward "a unified telecommunication network," but toward a "continued polarization process" around the two space powers: the Soviet Union and the United States. He enumerated the flaws that several jurists from the Soviet Union and other socialist countries had identified in COMSAT and Intelsat: the "peculiar intertwining of the private and state character"; the designation of the United States corporation COMSAT as the manager for Intelsat; and the absolute majority then held by the United States representatives on the governing committee of the Intelsat consortium.[80]

The Soviet Union was very critical of the initiation and development of COMSAT and Intelsat. It objected to the provision that Intelsat membership was open only to members of the ITU, a condition that, it insisted, put the claim to universality in doubt and violated "the principle of sovereign equality of states." The Soviet jurist V. S. Vereschetin commented on COMSAT and Intelsat's attempt to prohibit Intelsat members from participating "in the establishment or use of other international space communications systems." (This proviso was presumably designed to prevent the rise of a rival "global satellite communications" system. In an interesting way, it echoes the earlier efforts by the British Marconi Company and by the four-company cartel in South America—led by General Electric—to use nonreciprocity to keep out competitors.) Vereschetin felt that the prohibition contradicted the Outer Space Treaty specification that outer space "shall be free for exploration and use by all states without discrimination of any kind on a basis of equality."[81] It was the Soviet view in 1969 that "a genuinely universal [satellite] communications system" could develop only through a "spirit of cooperation on the part of all states and rejection of attempts to set up in an international communications organization monopoly of one state or group of states."[82]

In any event, the effort to prevent Intelsat members from participating in other space communications systems was not successful. European countries wanted no such restriction; a compromise provided that should any public international satellite system be set up by Intelsat members outside of Intelsat, it would be such as to avoid "significant economic harm" to Intelsat.[83]

The other challenges to Intelsat procedures were refuted by, among others, Richard R. Colino, COMSAT's legal counsel, speaking at the colloquium of the IISL in 1970 (in the 1980s, he became, for a time, Intelsat's director general). Colino pointed out that 137 nations were members of the ITU and that those few that were not were of "minuscule" importance in international communications (they included China, North Vietnam, North Korea, and the German Democratic Republic, which helps explain the Soviet position). Intelsat's voting procedures, which gave a dominant role to the United States, were also defended: "The principle that a member's [financial] investment in the organization and its vote in its governing body should be related to its use of the system is probably the most objective criterion which can be employed in an organization such as Intelsat."[84]

At the Fifty-second Conference of the International Law Association, held in Helsinki in 1970, the Soviet representative reported on the development of Intersputnik, the Soviet-inspired multinational communication-satellite system. Intersputnik, in effect, put an end to the hopes of Intelsat to become the single global system. But despite the widening gulf between the United States and the Soviet Union on the Intelsat issue, space-law theorists continued speculating hopefully about the possibility of a world system, citing their belief that in deliberating on the rules to govern the use of telecommunication satellites, "the jurist must bear in mind the high destiny of man in his relations with fellowmen of all the planet."[85] A compilation summarizing the first fifteen years of the proceedings of the International Institute of Space Law took note of IISL–sponsored discussions about the possibility of creating a United Nations space agency that could be made responsible for administering space telecommunications as well as other space activities. A Belgian jurist, Julian G. Verplaetse, told one of the colloquium sessions of the imperative need to create a world control system: Intelsat and Intersputnik could not "compel the universal interest of the world body politic"; there should be "a new authority for outer space telecommunications."[86]

It was unlikely that the controversy over the direct broadcast satellite could have been resolved through the mechanism of a supranational network, nor that Intelsat would have been an acceptable candidate for the task, although it had its supporters. For example, M. S. Geston suggested that "a commercial [system], controlled by votes proportional to investment," would be a promising approach.[87] If it were to be Intelsat, however, it would have to distance itself from the United States presence, which would be difficult to do "without destroying . . . [Intelsat's] ability to draw upon the technology and support of the leader in space technology."[88] In the event, Intelsat eventually faced competition not only from Intersputnik, but also from Western Europe, as countries came together to seek "space independence" through the European Space Agency (ESA).[89]

As COMSAT and Intelsat expanded and U. S. technology gained ground, it became apparent to other nations that the United States, or corporations within the United States, would sooner or later develop the capability to broadcast by satellite commercial or government television signals to home receivers in other countries. It had not gone unnoticed that the United States had sought, unsuccessfully, at an ITU conference in 1971 to get the members of the ITU to agree to allocate for satellite broadcasting a segment of the frequency spectrum below UHF, which would have permitted reception of DBS by existing home television sets. The ITU did allocate frequencies for DBS in 1971 but in the higher range, where there was less congestion and no home television use was technically possible.[90] As we know, the issue was finally joined in the 1972 session of the United Nations, when by vote of 101 to 1 the General Assembly asked COPUOS to develop guidelines for the regulation of such broadcasting with a view to formulating a treaty.

The DBS Debate in the United Nations

The substance of the legal debate over DBS in the years after 1972 was influenced by key decisions made in three international bodies—the ITU, UNESCO, and the United Nations—as well as by the larger political context, including the continuing détente between the Soviet Union and the United States, the resurgence of Japan and Western Europe, and the increasing solidarity of a maturing Third World.

In 1971 the ITU had for the first time made frequency assignments for the broadcasting satellite services, although not specifically for DBS. It also had adopted a new regulation, 428A, which requires coordination between broadcasting nations and receiving nations prior to the activation of a satellite broadcast. The regulation states that "in devising the characteristics of a space station in the Broadcasting Satellite Service, all technical means available shall be used to reduce, to the maximum extent practicable, the radiation over the territory of other countries unless an agreement has been previously reached with such countries."[91]

It was an action of primary importance to the issue of prior consent, and one commentator suggested that the ITU members would not have agreed to allocate frequency bands for satellite broadcast if this provision had not been agreed to at the same time.[92] Chapter 5 discusses this development and its effect on the DBS debate.

In another international arena in 1972, UNESCO concluded its long deliberations on the issues surrounding direct broadcast by satellite with the adoption of the Declaration of Guiding Principles on the Use of Satellite Broadcasting for the Free Flow of Information, the Spread of Education and Greater Cultural Exchange.[93] Like most UNESCO agreements, the declaration was not binding but was set down as a form of international consensus calling on states to reach or promote prior agreements before making satellite broadcasts into other states, and proposing the development of restrictions on content.[94] Later that year, the topic of regulations came before the United Nations itself as a result of the Soviet Union's proposal for a treaty to specify limits on the content of television broadcasting by satellites and to prohibit the sending of television broadcast signals via satellite into a country without that country's consent. The General Assembly did not accept the Soviet proposals but decided that there was a need for principles governing DBS and voted to assign COPUOS the task of developing them as a basis for a possible treaty. Although the United States voted against this, it subsequently joined without demurral in the COPUOS deliberations on the issue, deliberations that were to continue for ten years.

The early part of the 1970s was a period of détente, an apparent relaxation of the hostility and distrust between the United States and the Soviet Union. This was attributable both to a growing interest in the United States in expanding trade relations with the socialist countries and to the growing evidence of the Soviet Union's nuclear capabilities, which made it less willing to passively accept the United States' long-established encirclement policies and brought about a recognition in the United States of the desirability of finding mutually advantageous ways of coexisting. Perhaps the high point of the détente cycle came in 1975, with the joint manned space flight and the signing of the Helsinki Accords. Scientific contacts, especially in space-related activities, continued throughout the decade.

The maturation of the European Economic Community and the Council of Europe (its parliamentary arm in Strasbourg), as well as the economic resurgence of Japan and the Federal Republic of Germany, also had significant effects in altering the United States' ability to influence world politics. In the area of communications, the satellite successes of the Europeans and the Japanese introduced a competitive alternative to U.S.–based corporations that had enjoyed commercial and technological supremacy since the end of World War II.

Finally, many of the nonaligned countries, where poverty, disease, illiteracy, and even famine continued undiminished, were pressing forward with their efforts to change traditional patterns of world wealth and power distribution, working through the United Nations General Assembly and UNESCO toward ever more ambitious goals. The demand of the delegates to the UNESCO meeting held in Nairobi in 1976 for a "new world information order" was as unpopular among Western media interests as had been the declaration calling for DBS prior-consent guidelines, adopted in 1972. An unprecedented display of Third World independence over the issue of satellite communications occurred in 1975, when a group of equatorial countries (taking note, no doubt, of OPEC's success in forming a united front against the First World) issued the Bogotá Declaration. In it, the signatories asserted their sovereignty over the geostationary orbit—a bit of international legal bravado that gave new life to the effort to decide where air-space sovereignty ends and outer space ("the province of all mankind") begins. Still another setback for the United States in its continuing effort to fend off the smaller nations' claims to such rights as prior consent, communications development aid, and new world information orders was the decision in 1978 to enlarge the membership of COPUOS from thirty-seven to forty-seven, which, given the make-up of the United Nations, could only increase the number of votes favoring the interests of the smaller countries.

The 1970s were also marked by steady gains in space-communications technology, in part because of the critical importance of this technology to the growing surveillance and war-making machinery of the United States and NATO, and in part because of the burgeoning communications demands of the rapidly expanding capitalist world market. The first DBS application, using community receivers, was demonstrated by NASA in India in 1975. A similar program, designed to reach populations in the sub-Arctic, was initiated in Canada in 1975. In 1976 was established the International Maritime Satellite Organization, with the aim of developing satellites and ground stations as the collective property of all participants (its member states include both the United States and the Soviet Union). In 1978, the United States and the Soviet Union converted their "hot line" from undersea cable to space satellite, and NASA launched a DBS satellite under contract with Japan. By 1979, Intelsat could report that it had 12 international communications satellites in synchronous orbit, with 203 earth stations in 97 countries. It was also providing satellite capacity for national domestic services in 15 countries. It should be kept in mind that the bulk of Intelsat's transmissions are business services—for example, electronic fund transfers ($5 trillion in 1983) and telephone calls (1 billion in 1983). Television transmissions have formed but a small part of its annual business, and none have been direct broadcast by satellite (DBS) but broadcasts transmitted by prior agreement between various national television services. Spokespersons for Intersputnik reported ten member countries in 1979, with nine earth stations and a number of communication satellites in geostationary orbit above the Atlantic and Indian oceans, sufficient to "make it possible to ensure communication practically between all countries."[95]

Against this background evolved the debate over how to control DBS, which has been carried on in a number of arenas and has engaged the talents of diplomats, negotiators, international lawyers, and social scientists. The Legal Subcommittee of COPUOS and, especially, its Working Group on DBS have been the major forums, holding annual meetings and presenting detailed reports to the General Assembly. Law journals and committees of the American Bar Association have provided a continuing flow of DBS analysis. At the international level, the annual colloquia of the IISL have explored the issue and made recommendations. They have proved to be a major source of information about the views of lawyers from the Soviet Union and other socialist countries, who between 1973 and 1979 presented fifteen monographs on various aspects of the DBS controversy.[96]

In the early years of the COPUOS deliberations on the issue, the Soviet draft convention formed the basis of the discussions. Of particular interest to Western participants in the controversy was the Soviet proposal that violations of the proposed convention would allow the offended state to employ "the means at its disposal to counteract illegal television broadcasting of which it is the object, not only in its own territory but also in outer space and other areas beyond the limits of the national jurisdiction of any state."[97]

In the face of strong criticism, the Soviet Union softened this proposal in 1974, making it clear that the possibility of counteracting a satellite signal was to be accomplished by means "recognized as legal under international law," not by military retaliation.[98] However, it held to the prior-consent proviso and to the proposals for program-content prohibition of "ideas of war, militarism, national and racial hatred and enmity between people, as well as material which is immoral or instigating in nature or is otherwise aimed at interfering in the domestic affairs or foreign policy of other states."[99]

In 1973, Sweden and Canada offered a different version of a possible DBS treaty, retaining the requirement of prior consent and giving attention to the term "free flow of information," which had been ignored by the Soviet draft. The new version affirmed that direct broadcast by satellite "shall be conducted on the basis of respect for the principles of sovereignty of states, non-intervention, and equality and in the interest of promoting the free flow of communications."[100]

Between 1974 and 1979, prodded by a sometimes impatient General Assembly, the member states of COPUOS, adhering to consensus procedures (as they had from the beginning), arrived at unanimous agreement on nine of the proposed principles affecting DBS. They could not agree on the remaining three: prior consent, prohibitions on broadcasting content, and possible recourse to state action aimed at stopping unwanted broadcasts. The list of principles on which they did agree is a significant indicator of the extent to which differing viewpoints could be harmonized, including a degree of compromise between the United States and the Soviet Union—the central opponents in the negotiation process. The nine points of agreement were:[101]

1. Purpose and objectives. It was agreed that among other things, DBS was to be used to strengthen international understanding and cooperation in maintaining peace and security and to promote dissemination and exchange of knowledge and information.
2. Applicability of international law. The Outer Space Treaty, the United Nations Charter, and international law were determined to be applicable to DBS. However, proposals to cite other specific sources of international law, such as the ITU convention and the UNESCO declaration of 1972, were not agreed to, nor was there agreement on in-

cluding related legal instruments that deal with fostering friendly relations among nations.

3. Rights and benefits. It was agreed that all states have an equal right to utilize DBS and to authorize its use by entities under their jurisdiction.
4. International cooperation. It was agreed that international cooperation was desirable, at regional as well as international levels.
5. State responsibility. This agreement reiterated the provisions of Article VI of the Outer Space Treaty, which holds that states are responsible for DBS activities carried out under their jurisdiction.
6. Duty and right to consult. There was agreement that an aggrieved state might seek consultation with the broadcasting state concerning "any matter" arising out of DBS penetration. Significantly, agreement on this point was made contingent on the unresolved questions of prior consent and content prohibitions, which if adopted would nullify this principle.
7. Peaceful settlement of disputes. It was agreed that peaceful mechanisms should be used to settle disputes, but there was no provision for binding arbitration.
8. Copyright and neighboring rights. It was agreed that existing international copyright instruments were applicable to DBS; in connection with the application of copyright requirements to developing countries, the negotiators proposed that such countries receive special consideration (as they do under most law concerning protection of intellectual property).
9. Notification to the United Nations. It was agreed that states engaging in international direct television broadcasting by satellite should so inform the Secretary General of the United Nations (as with other outer-space activities).

Despite the advances represented by these agreements, COPUOS was unable to achieve its goal of drafting a set of DBS regulatory principles for submission to the United Nations. United States opposition to the remaining central issues—prior consent, program-content restrictions, and possible recourse to stop unwanted broadcasts—stymied further progress. For some, this could be viewed as a victory for the U.S. position, since as long as these issues remained unresolved, technological and practical advances could continue to go forward without restraint. But as we have seen, proponents of prior consent were winning ground in other arenas—specifically, in the ITU (Regulation 428A), in UNESCO (the new world information order), and in elaborating the full meaning of the applicable international law (the United Nations Charter, the Outer Space Treaty, and such), thus mooting jurisdiction of COPUOS in the matter.

5

The U.S. Position in the DBS Debate

"Free Flow of Information"

One of the central arguments for the U.S. position on DBS is that the free movement of information is an objective force that "injures no nation" and that an "open marketplace of ideas and information is essential to the well-being of the international community." A second tenet frequently cited is the belief that "our own fundamental law" (that is, the constitutional protection of free speech) prohibits the United States from becoming party to any international agreement that would restrict "the right of free expression." These arguments were set forth in 1973 by F. S. Ruddy, then deputy general counsel for the U.S. Information Agency (USIA), speaking at the sixteenth IISL colloquium.[1]

Both of these arguments have been weighed and elaborated by American legal scholars, encouraged to do so by foundation grants, law-school study programs, and the State Department, which used its program of external research to bring together a number of American legal experts and communications sociologists at a conference—"The Free Exchange of Information and Ideas and the Integrity of National Cultures"—held in Airlie, Virginia, in February 1974.[2] A sense of the State Department's central concerns is reflected in the four questions posed to the experts in the "Draft Work Statement" prepared for the conference:

1. How well established in principle and in practice internationally is the right to receive and impart information across national boundaries?
2. Does the right to receive information and ideas apply more to certain types of information and ideas than to others—for example, is the right of a citizen of any country to receive cosmetic advertising from abroad equivalent to his or her right to receive debates by the United Nations Security Council?
3. Does the practice of governments that impose controls on the volume of imported information material, such as films and television programs, violate the principle of "free flow"?
4. What validity is there to the argument of some governments that they must control foreign media that are assumed to have an intense impact, such as television, as contrasted with media that have less impact, such as shortwave radio?[3]

Before examining current views of "free flow" as a rationale for the official U.S. opposition to prior consent, it may be useful to summarize the history of the term and of the conditions it presupposes. It began to appear in international debates during the opening stages of the cold war, following Churchill's "iron curtain" speech in Missouri. Thereafter, it became a key slogan in the human-rights debates in the United Nations and UNESCO, debates that focused on the fundamental economic and political differences between the Soviet Union and the United States and that culminated in 1950 in such actions as the drafting of General Assembly Resolution on Freedom of Information and the creation of the Freedom of Information Committee of UNESCO. Among those participating in these debates were representatives of the U.S. publishing and broadcasting industries, which were seeking to affirm their longstanding access to and distribution of news through their worldwide news agencies and networks.[4] The phrase "free flow" was also widely used in the promotion of the Radio Free Europe campaign. In the view of the Soviet Union and other socialist countries, the phrase came to be associated with the establishment of Radio Free Europe and Radio Liberty. They regarded these stations as a part of the cold war, "in which, in contrast to hot war, it is not the cannons that speak but subversive radio stations, newspapers, magazines and other mass information and propaganda media."[5] Y. N. Zasursky, writing in Prague in 1974, asserted that the freedom of the "free flow of information" was, in fact, not individual freedom but freedom for the media conglomerates or state broadcasting agencies of the Western countries. The information part of the slogan, he asserted, was misinformation, pointing to Lord Beaverbrook's statement that he published his newspapers for propaganda purposes. Zasursky proposed that "free flow" be renamed "electronic imperialism."[6]

Keeping in mind the differing interpretations of "free flow," we can consider its importance and value to the theorists who were called on to advise the State Department at its conference in 1974.

In his contribution, John L. Hargrove, then acting head of the American Society of International Law, offered a sophisticated rationale for rejecting the established principle that the state is an irreducible entity in international law. Like Zasursky, Hargrove borrowed his imagery from the politics of world commerce. To Zasursky, "free flow of information" was a disguise for "electronic imperialism"; to Hargrove, the proposed restraint of information flow was a form of "cultural protectionism." Hargrove pointed out that the idea of negotiating an international agreement to legalize such "cultural protectionism" was based on acceptance by the international community of the inviolability of each state. That, he argued, was not a sufficient cause for such protection, especially if what was being protected was a "ruling elite"[7] or "if there is reason to doubt the representative character of the national governmental apparatus."[8] The object of "the evolving international community" should be "to develop its own capacity to hold local authorities accountable as fiduciaries for their own populations."[9] Recognizing the impracticality of his prescription, Hargrove nonetheless recommended against any international treaty or agreement to limit communication or other cultural interchange, since such "cultural protectionism" would have the effect of "abetting the unreasonable infringement of legitimate community interests by national authorities."[10]

This statement against prior consent is significant on two counts. First, it makes explicit what is often alleged to be one of the goals of international shortwave broadcasting: the undermining of the established authority of a state when that authority is deemed by the broadcasting state to be unrepresentative. If Radio Free Europe's broadcasts are considered by their sender (the United States) to be in the interests of the people of Poland, as distinguished from an elite perceived to be ruling Poland, Hargrove's views would legitimize this type of foreign influence. Hargrove did distinguish between what he called "cultural exchange" and what he recognized as broadcasts from outside a state that threaten "force" (his quotation marks) or urge others to bring about the state's forcible overthrow. These latter kinds of communication, he felt, were already prohibited by the United Nations Charter, a significant admission that received no further comment. Second, Hargrove's argument raises a key issue: the validity of the state as a reflection of its citizens. A thorough examination of the issue would take us too far from our topic, but it should not be overlooked as a consideration in any discussion of information resources and their control.

Because the nation-state is the most powerful form of political organization we know, we often tend to regard it as an unchangeable given. The possibility of higher, more complex, and more comprehensive forms of political organization must not be foreclosed, although it must be recognized that the struggles between the holders of power and the powerless will continue even after larger political entities have been established and will permeate the stages of development needed to reach them. Hargrove is representative of those who would undermine "statism" by pitting one state against another; others would build on the structures of states to reach new forms of international cooperation; still others would introduce the theories of nineteenth-century Marxism and twentieth-century Leninism, both of which anticipate the replacement of capitalist nation-states with international organizational forms of communism.

Hargrove's opposition to any international invocation of "cultural protectionism," or prior consent, is a more extreme position than that taken by most American jurists, who do not generally question the paramountcy of state sovereignty. Some have questioned it, as in this comment that "the sovereignty of states is not paramount to other rights under international law, including the right of scientific research, the free flow of information, and the use of outer space for peaceful purposes."[11] Sig Mickelson, a former executive of CBS and former president of Radio Free Europe and Radio Liberty, has complained about "the Soviet Union's direct attempt to impose the theory of complete state sovereignty over all forms of communication."[12] But most professors and practitioners of international law have not aligned themselves with the official U.S. "free-flow" position, and several have suggested that it is less than legally acceptable. At the very start of the DBS controversy, in 1969, Paul L. Laskin wrote a Twentieth Century Fund report proposing that states that might broadcast via satellite "adopt the simple principle that they will not use satellites for the deliberate broadcasting of programs to any country without the recipient's consent."[13]

Paul Laskin and Abram Chayes were among the experts consulted by the State Department at its conference in 1974, and they were also leading participants in a conference of experts from many countries that was held that year at the Rockefeller Foundation's conference villa in Bellagio, Italy. Their analyses of the international-law aspects of DBS have been published in several books and journals and provide an extended view of the history and complexity of the U.S. position.[14] By 1974, as already noted, the United States officially appeared to be no longer opposing the idea of a declaration of United Nations principles regarding television DBS. It had submitted to COPUOS its own draft, which stated that DTBS (the *t* stands for *television,* the first time this abbreviation appeared) "should be carried out in a manner compatible with the maintenance of international peace and security with a view to enhancing cooperation, mutual understanding, and friendly relations among all states and peoples."[15] The draft suggested that international DTBS encourage "the free and open exchange of information and ideas," but precluded any kind of prior consent. Laskin and Chayes have developed their argument by citing the several international declarations in support of the "free flow of information and ideas," including the United Nations Charter, which provides that member states are obliged to promote "universal respect for and observance of human rights and fundamental freedoms," and the 1948 Universal Declaration of Human Rights, which spells out some of these rights. Article 19 of the Universal Declaration of Human Rights states that "everyone has the right to freedom of opinion and expression; this right includes freedom to hold opinions without interference and to seek, receive and import information and ideas through any media and regardless of frontiers." Not a treaty, the declaration has nevertheless gained wide respect as an authoritative elaboration of United Nations goals and has been reaffirmed in a number of international human-rights instruments, as will be detailed later. Chayes and Laskin asserted that "any regulatory scheme for international satellite broadcasting is to facilitate and enlarge the flow of information and ideas, not to stifle it"; it must also take into account "the value of cultural diversity for the world at large."[16] At the same time, they cited the principle of state sovereignty: "that every nation attaches [importance] to the right to determine for itself, at least in a general way, the scope and character of the television services available to its people."[17]

Measuring the U.S. position against these three principles, they concluded that "it fails to give proper weight to the interest in diversity and the desire of States to determine for themselves the character of their national television services." They added:

> One may ask, for example, what the reaction of the United States would be if the viability of the structure of American television were threatened by foreign

> satellite broadcasts. The United States would undoubtedly assert the right to take protective action to ensure the continuance of a domestic television service adequate by its own standard. From the point of view of countries with little need or desire to engage in international direct broadcasting (many of them undeniably devoted to the principle of freedom of information), the consequences of the minimalist position put forward by the United States look in practice very much like foreign intrusion and interference.[18]

Chayes and Laskin found merit in the draft proposed by Canada and Sweden. This draft supported prior consent and granted a receiving country the right to participate in discussions about radiation of its territory, a provision that Chayes and Laskin felt would adequately protect national autonomy in television matters. But can prior consent be reconciled with "the principle of free flow of information and ideas"? They pointed to the longstanding de facto power of a state to control any information coming into it from foreign sources: mail, telegraph, print media, recordings, tapes, television signals received via centrally controlled receivers. They included even shortwave-radio signals, which have long been combated by jamming: "While jamming is not an internationally approved practice, especially by nations with a strong tradition of free speech, it has never been considered a violation of an international legal obligation."[19]

Concluding their commentary on the issue, Chayes and Laskin proposed some refinements to the prior-consent draft of Canada and Sweden:

> First, a right of consent should be exercised only in accordance with the principles and norms that have been developed, and are still being developed, by the international community, including especially the principle of free flow of information. Second, satellite broadcasting should be governed by broad consensual arrangements negotiated between transmitting and receiving States which set out the terms and conditions of the enterprise.[20]

Thomas Buergenthal, another professor of law who spoke at the State Department seminar, attached importance to the fact that the Universal Declaration of Human Rights, while it protects the right to receive and impart information, provides that this protection may be legitimately restricted for reasons of "public order" and "public welfare" (Article 29[2]). Looking at the ongoing prior-consent debate among states, Buergenthal felt that "few if any states" were challenging "the principle of freedom of information as such" but were seeking prior consent "as a way to ensure the enforcement of those limitations on freedom of information that a state deemed necessary to protect or promote its public policy, be it in the political, economic, social or cultural sphere."[21] Buergenthal is one of the few who have commented on the use of DBS for advertising. He felt that states could justly ban unwanted advertisements delivered by DBS, since "under international law each state is free, in the absence of specific international agreements to the contrary, to bar aliens from engaging in commercial activities within its territory." He reasoned that "advertising whose sole purpose is to propose a commercial transaction could be equated to commercial activities in general" and thus could be "subject to the exclusive regulatory authority of each individual state."[22] Buergenthal suggested that other states fear that "the U.S. will ride roughshod over the public policy of all other states," a fear he felt was fed by Americans who sometimes expressed the view that "the principle of freedom of information is absolute or close to it, which of course it is not, even under our own laws."[23] States that were seeking protection from unwanted signals might, he felt, legitimately insist on "more balanced programs," "equal time" or the "right to reply," and time limitations, but he concluded that "an absolute right of prior consent would negate the very existence of the international right to receive and impart information across national boundaries."[24]

In 1978, the Brookings Institution published a study that focused on "conflict resolution" in three areas: outer space, oceans, and weather—all of which "elude the kinds of local and national regulations applied to commonly used resources on land."[25] The familiar Lasswell formula was invoked: "Who gets what, when, and how in the nonland areas can affect the security, health, economic welfare and power of ever larger portions of the earth's population."[26] After reviewing the DBS controversy, the authors rejected a "prior consent regime," which would be "prior censorship beyond that practiced by the United States government for domestic programs" and "would probably contravene the First Amendment to the Constitution."[27] In the absence of such a regime, they felt that "the United States would probably continue to broadcast internationally, and other countries would have to resort to jamming to prevent unwanted broadcasts."[28] They proposed as an "alternative" what they called an "Agreement Regime": "A general international convention" that would "stipulate that broadcasting via satellite into foreign countries should be with the agreement of the involved countries."[29] Untroubled by the apparent contradiction in their position, they stated: "We are convinced that uncompromising U.S. insistence on an absolutely open regime for satellite broadcasting would work contrary to present arrangements and would impede extension of any eventual capabilities for direct broadcasting."[30] The authors commented on a policy difference between "communications professionals in government international broadcasting circles" and persons working within the USIA. They felt that the professionals seemed to acknowledge the marginality of "intrusive [satellite] broadcasting," but not so the USIA, which "for many years . . . has, in its Congressional testimony, left open the door to satellite broadcasting."[31] The authors alluded to the question of advertising by DBS, believing it to be an unlikely development for reasons of cost, limited audiences, and the ready availability of other advertising methods.[32]

A very useful monograph by Leo Gross, professor emeritus of international law at the Fletcher School of Law and Diplomacy, summarizes the conflict between state sovereighty and "free flow."[33] Gross cited the *Lotus* case, decided in 1927 by the Permanent Court of International Justice (predecessor of the World Court), in which the court said that international law emanates from the freely taken actions of states, in the form of conventions or usage, and that "restrictions upon the independence of States cannot therefore be presumed."[34] Gross warned against "the mistaken notion that the limitations that international law inevitably imposes upon States are somehow contrary to, or incompatible with, their sovereignty." On the contrary, "compliance with the law is inseparable from the very notion of international law."[35]

Another basic principle that Gross deemed relevant is that states "may not invoke provisions of their internal law as justification for their failure to perform their obligations under treaties or under customary international law."[36] He also called attention to the principle known as domestic jurisdiction, reinforced by the principle of nonintervention, which provides that in the absence of applicable international law, "states are free to exercise their sovereignty within their territory."

Gross gave close attention to the argument that United Nations resolutions, such as the all-important Universal Declaration of Human Rights, are international law and thus binding on the member states of the United Nations. He concluded that they are "merely statements of objectives and purposes" and do not become legally binding unless or until states convert them into treaty form. To those who have argued that the Universal Declaration of Human Rights has become customary international law, he pointed out the fundamental requirements necessary for a measure to achieve such acceptance: it must be observed, and the observing states must believe that it is obligatory to do so.[37] He noted that United Nations members, after two decades of work, had transformed the Universal Declaration of Human Rights into international law in the form of two documents—the International Covenant on Economic, Social and Cultural Rights and the International Covenant on Civil and Political Rights—both of which were ratified by enough states to go into force in 1976. The former refers to the rights of individuals with regard to cultural and scientific pursuits, and the latter deals with freedom of information, as follows:

> *Article 19*
>
> 1. Everyone shall have the right to hold opinions without interference.
>
> 2. Everyone shall have the right to freedom of expression; this right shall include freedom to seek, receive and impart information and ideas of all kinds, regardless of frontiers, either orally, in writing or in print, in the form of art, or through any other media of his choice.
>
> 3. The exercise of the rights provided for in paragraph 2 of this Article carries with it special duties and responsibilities. It may therefore be subject to certain restrictions, but these shall only be such as are provided by law and are necessary: (a) for respect of the rights or reputations of others; and (b) for the protection of national security or of public order . . . or of public health or morals.
>
> *Article 20*
>
> 1. Any propaganda for war shall be prohibited by law.
>
> 2. Any advocacy of national, racial or religious hatred that constitutes incitement to discrimination, hostility or violence shall be prohibited by law.[38]

Both covenants were signed by the Soviet Union; both covenants were signed by President Carter on October 5, 1977, but have not been ratified by the Senate. Gross observed that the International Covenant on Civil and Political Rights provides a way for the United States to make the right to freedom of information part of a legally binding treaty, although it is "not clear how far, if at all, it confers rights upon the *media themselves* [Gross's italics]."[39] He concluded: "If the United States takes the freedom of information and other rights seriously and desires to play a significant role in their application, it must ratify the Covenant."[40]

One further aspect of the "free-flow" versus prior-consent argument should be noted. It has to do with the uncertainty surrounding the use of DBS for shortwave-radio transmissions. Several authorities have brought up this question, suggesting that any restrictions that might be adopted by treaty to control unwanted television broadcasts via satellite might spill over into other forms of telecommunication, whether by satellite or not. One American expert suggested that the proposed DBS restrictions might "impede or inhibit the free flow of information by other media where legal, and thereby reduce international understanding and cooperation between the peoples of the world."[41] Another asked, "What are the implications [of a consent principle] for other communications media. . .? Do we really want to open up the use of such a principle as a precedent that could be invoked to stem the free flow of information and ideas?"[42] A German jurist, expressing opposition to "rigid . . . regulations . . . as proposed by the Soviet Union," also raised the fear that if television broadcasts by satellite were subject to prior consent, it might spread out to cover other media: "We fear . . . that one can very easily transform such guidelines to terrestrial TV, to shortwave, longwave, mediumwave radio and all kinds of communications."[43]

In summary, what American diplomats and sociologists of communication have called "free flow" or "freedom of information" has from the beginning been a central element of the government's official defense against any form of prior consent. "Free flow" is regarded as a formal "human right," independent of a social or political context, and is frequently proclaimed in absolutist terms, despite longstanding and well-recognized internal limits on information collection and distribution. It is held to on ideological grounds, and in part because of a practical fear that if adopted, prior consent might be extended to apply not only to television signals, but also to the more traditional forms of mass communication across national borders and, possibly, even to data flow. However, when the "free-flow" argument was subjected to rigorous scrutiny by legal scholars, mainly in the United States, it did not win their endorsement. It was seen to be in conflict with the principle of state sovereignty, with practices in the United States, and, possibly, with international law and the principles of the United Nations.

The First Amendment

If American international-law experts were unwilling to defend the concept of "free flow" as officially propounded by their government, they were even less able to give credibility to the argument that an international prior-consent agreement would nullify the First Amendment of the Constitution. The position of the United States was expressed in 1973 by a representative of the State Department: "For reasons of our own fundamental law we are without power to agree to any instrument which barters away the right of free expression."[44]

The American tradition of free speech was a major theme of Harry Olsson, who was general attorney for CBS at the time that he participated (along with top-level representatives of NBC and ABC) in the State Department's conference in 1974. In that year, he spoke at a forum on the direct broadcast satellite sponsored by the Columbia University School of Journalism.[45] He insisted that restrictions not be placed on broadcasting across international borders—whether of news, entertainment, or commercial advertising—and felt that any agreement with other states on "mutual cooperation" over such programs would mean "compromising our basic principles." Olsson argued that "if we abandon the principle of free speech for the sake of foreigners, are we not necessarily going to lose the principle . . . in the domestic area as well?"[46] He called particular attention to a Soviet proposal that "advertising and other commercial material may be carried out only on the basis of specific agreements" between the states concerned, an assertion that he described as "a direct attack on our system of broadcasting. It singles us out for discrimination among the types of broadcasting systems in the world."[47] Later in the discussion, he said:

> What we are talking about here is the power of governments to step between those who broadcast and the people who want to receive the broadcast—the power of governments which want to keep information or even entertainment from these people. They want to make it impossible for people to make the choice as to whether they will listen or not.[48]

Olsson's statement echoed the position taken by Frank Stanton, vice president of CBS and an official adviser to the USIA. Stanton was reported as wishing to expose the issue to the public "when he saw—correctly—that the State Department was trying to avoid a head-on confrontation in the diplomatic arena."[49] Stanton had argued earlier that "the fundamental principle of free speech is at stake," and "the United States must do all within its power to block the path to international censorship."[50] In a speech given on October 4, 1974, Stanton made the same point: "I do not see how our government, given our constitution, can possibly enter into any agreement in which the rights of Americans to speak to whomever they please are bartered away."[51]

In hearings before a Senate subcommittee in 1976, Senator George McGovern rejected prior consent as "internationally approved censorship."[52] John E. Reinhardt, director of the USIA, said at the same hearings: "As we believe that free speech is democracy's chief weapon against tyranny, so do the leaders of the totalitarian states believe that control of free speech is a vital weapon in preserving their structure."[53]

A full examination of the alleged primacy of the First Amendment in the DBS controversy was undertaken in 1976 by Monroe E. Price, professor of law at UCLA.[54] Abram Chayes had argued the issue at the Columbia University forum on DBS. To Olsson's comments that the prohibition of commercials was an infringement on free speech, Chayes had responded by noting that FCC regulations establish limits on commercials that are designed for broadcast to foreign countries and prohibit the Public Broadcasting Service from carrying commercials, "without any involvement of the principle of freedom of information or the First Amendment."[55] Indeed, he pointed to "a whole series of defenses against speech being imposed on unwilling listeners."[56] After an exhaustive analysis of the matter, Price came to similar conclusions, citing compelling precedents that erode the U.S. government's claims that First Amendment imperatives preclude any prior-consent limitations. Price called attention to the regulations dealing with international mail, which are observed by the United States by virtue of its membership in the Universal Postal Union. The regulations prohibit the U.S. Postal Service from handling certain kinds of material, including mail that is prohibited in the receiving country.[57] Quoting from the *U.S. Postal Service Manual,* Price listed some of the prohibitions that have been set by other countries and that must be observed by a country sending mail into those countries: Argentina prohibited "communist propaganda"; Australia prohibits matter that unduly emphasizes "crime, horror or sex"; the Soviet Union prohibits toys and games "of a militaristic nature"; several countries restrict the entry of "exposed motion picture film." That this arrangement goes unchallenged is evidence "that the U.S. can constitutionally be party to an international arrangement for the exchange of information where one party-state reserves the right to censor."[58]

Another important precedent that Price identified was the established licensing and regulatory authority of the FCC, by which it regulates radio and television, including the control of "unrestricted competition" (from cable television). Price pointed to Supreme Court decisions upholding the FCC's extensive jurisdiction and citing language from the original legislation of 1934, which provided that the act should "apply to all interstate and foreign communication by wire or radio . . . which originates and/or is received within the United States." In speculating about the FCC's relation to any future prior-consent treaty, he expressed the belief that it would probably see itself as the proper agency to assert jurisdiction over direct broadcast signals unless specifically barred by Congress or treaty language.[59]

What about obligations to receiving states? Could the FCC assume functions equivalent to those of the Postal Service and screen outgoing programs of DBS to enforce "consent standards of a receiving state"? If the FCC had such extensive jurisdiction, how might it approach those consent requirements without infringing on First Amendment rights? Price skirted this problem and proposed instead that enforcement of any such restrictions be left to recipient nations (leaving open the question of how this might be accomplished if a sending country refused to comply). Thus free-speech rights in the United States would not be restrained by a U.S. control agency, beyond the limitations that already exist—for example, fairness doctrine, legislation requiring that "news be objective," and the like.[60] Significantly, Price did not mention the precedents set by State Department control over scientific and cultural exchanges: for example, denial of visas to keep out peace activists, as was done before the June 12, 1982, disarmament march, or to exclude those with views that dissent from mainstream U.S. views, as has been accomplished in the case of invited foreign scholars, writers, and artists; control of passports to keep U.S. citizens from traveling to such countries as the president may designate; and current efforts to achieve "voluntary" self-censorship on the part of scientists in the United States at scheduled international science conferences. Also ignored by legal scholars, in their studies of the First Amendment and DBS, is the paradox in the position of the United States, which insists, on the one hand, on the paramountcy of "free flow" over the exercise of state sovereignty by other states and, on the other hand, on the paramountcy of its own state sovereignty over the exercise of prior consent by other states.

Throughout the long debate over prior consent for DBS has run the subtheme of restrictions on program content by prohibiting DBS from publicizing "ideas of war, militarism, national and racial hatred and enmity between peoples, which is aimed at interfering in the domestic affairs of other states or which undermines the foundations of the local civilization, culture, way of life, traditions or language."[61] This prohibition is taken from the original draft of DBS principles proposed in 1972 by the Soviet Union, although in the 1979 negotiating session, the Soviet Union was described as willing to give up any specification as to program content if the key prior-consent provisions remained, a position concurred in by Sweden and Canada.[62] It is possible that the Soviet Union felt that there was sufficient protection in existing content restrictions—for example, those in the Universal Declaration of Human Rights, which prohibits "propaganda for war" and "all advocacy of national, racial or religious hatred that constitutes incitement to discrimination, hostility or violence," a prohibition that is also embodied in the International Covenant on Civil and Political Rights, which was signed by the Soviet Union but not ratified by the United States.

The objections of the United States to any such content restrictions have been based on First Amendment rights, a position that Price examined carefully. He found precedent for content control in the FCC regulation that stipulates the licensing of international broadcasting only if the service provided "will reflect the culture of this country and . . . will promote international goodwill, understanding and cooperation."[63] In addition, he called attention to the de facto control over broadcasting to areas outside this country that is exercised by the U.S. government "by virtue of the Voice of America's virtual monopoly of frequency hours."[64] As further evidence of how FCC regulations set limits on content, Price cited (1) the FCC authority to prescribe the kinds of service that can be offered by the various classes of licensees, (2) the FCC prohibition on the airing of business messages by amateur radio stations (or political messages, although Price did not mention that), and (3) the FCC authority exercised in the 1930s to "deny a license to a facility in the United States whose purpose was to transmit material to . . . a Mexican station for retransmission into the United States."[65]

Price also alluded to the long-established FCC regulations that deny licenses to "any corporation of which any officer or director is an alien or of which more than one-fifth of the capital stock is owned of record or voted by aliens or their representatives or by a foreign government . . . or by any corporation organized under the laws of a foreign country."[66] Price argued that such a limitation clearly diminishes the diversity of information broadcast on radio and television in the United States. This provision is presumed to be constitutional on the grounds of national security (although it has not been tested), and it may be regarded as setting an important precedent for empowering the U.S. government to regulate direct broadcast messages "to protect national security."[67]

As for limiting specific areas of program content—such as racial or religious attacks, violence, obscenity, pornography, or use of narcotics—Price cited a number of FCC practices (and industry codes) that set standards that preclude such broadcasts in the United States.[68]

Price also considered the First Amendment in relation to possible governmental interference with broadcasts from other countries and found that "consistent with the first amendment, some classes of foreign broadcasts can be practically prevented from penetrating American borders."[69] He did not mention the decision of the Montreux International Telecommunications Convention of 1965, which authorizes members to suspend international telecommunications services or to stop information transmissions via telecommunications that are deemed dangerous to the security of a member state or contrary to its laws, public order, or decency.

Based on the evidence that he presented, Price concluded that the First Amendment "is consonant with an international scheme that classifies the spectrum available for direct broadcast by satellite so that only particular categories of programming could be used." Moreover, the First Amendment does not rule out an agreement that each nation could protect its own domestic television system from foreign competition. "Nor would the first amendment be inconsistent with international regulation of commercial messages." Some content restrictions would be consistent; others perhaps would not be: the mode of enforcement would be an important determinant.[70]

Other lawyers have interpreted the issue differently. Jennifer Freeman raised the First Amendment issue in an entirely different context, questioning the extent to which the free-speech provisions of the Constitution would even apply to an international DBS treaty. This is a controversial and unresolved question, although at least one decision answered it in the affirmative: "Neither the constitution nor any law passed in pursuance of it have any force in foreign territory unless in respect of our own citizens . . . ; and operation of the nation in such territory must be governed by treaties, international understandings and compacts and the principles of international law."[71]

The frailty of the First Amendment defense as justification for U.S. opposition to international control of DBS has become evident as the DBS negotiations have continued. Whether it is a sufficient defense to legitimize the claims on behalf of "free flow of information" depends, finally, on the very definition of freedom, still a source of contention in this divided world. But while the debate has proceeded, other complicating issues have arisen.

The ITU and Regulation 428A

Of particular importance in the controversy over the direct broadcast satellite has been the ITU policy known as Regulation 428A, which was enacted in 1971 as a part of the ITU regulatory provisions, which were ratified by the United States on July 14, 1972, and took effect on January 1, 1973. The far-reaching effects of Regulation 428A were first amplified by a Czech lawyer, Jan Busak, at the sixteenth annual IISL colloquium. To cope with spillover, which occurs when one nation's signal penetrates and thus interferes with the television service of another nation, Regulation 428A provides:

> In devising the characteristics of a space station in the broadcast satellite service, all technical means available shall be used to reduce, to the maximum extent practicable, the radiation of space stations in the broadcasting satellite service over the territory of other countries unless an agreement has been previously reached with such countries.[72]

Busak compared this rule with a longstanding ITU regulation affecting terrestrial radio. The regulation prohibits stations in one country from using a signal strength greater than that necessary for a national service of good quality, a rule designed to prevent interference with the signal of the domestic service of another country. He believed that the new principle in Regulation 428A was insufficient protection for countries desiring prior-consent protection against DBS and argued that additional principles should be adopted by the 143 member states of the ITU.

As the DBS negotiations went on, some states began to see Regulation 428A as a way out, suggesting that it was tantamount to prior consent and should or could be expanded to apply to broadcasting from satellites into home receivers. An analysis by two Canadian legal authorities cited the compromise draft on the DBS issue offered in 1979 at the COPUOS working session by Sweden and Canada. The draft provided that any direct television broadcasting by satellite "specifically directed at a foreign state, could be established only when it is not inconsistent with the provisions of the relevant instruments of the International Telecommunication Union,"[73] thus seeming to stipulate that if a state moved to set up a supranational service, it could proceed only in accordance with Regulation 428A. Most countries, however, continued to press for a set of COPUOS principles that would give a clearer statement of the principle of prior consent.

The ITU secretariat collaborated with the COPUOS working group during its long deliberations. In 1977, ITU Deputy Secretary General Richard E. Butler enumerated ITU actions in addition to Regulation 428A that were relevant to the COPUOS negotiations: first, the adoption of the principle of the right of ITU member states to equal access to the frequency bands that had been set aside for radio communication via satellite; second, the proviso that when a member state registers a frequency assignment with the ITU, it does not thereby gain permanent rights to that frequency; third, the decision to complete the DBS planning for Region 1 (Asia) and Region 3 (Europe and Africa) in 1977 and for Region 2 (North and South America) in the early 1980s; and, finally, the adoption of Provision 470V, which requires that DBS satellites be "fitted with devices to ensure immediate cessation of their radio emissions by telecommand, whenever such cessation is required" under ITU regulations.[74] It is noteworthy that the planning for Regions 1 and 3, which took effect on January 1, 1979, was made in accordance with Regulation 428A.[75] According to observers, the regional planning in Europe occurred at a surprising rate, with all affected countries agreeing to accept unintentional but unavoidable spillover signals from other countries' satellite-transmitted broadcasts. Chapman and Warren commented: "It is amazing that it proved possible to agree upon such a plan given the fact that the details of coverage areas and 'spillover' affected the sensitive interests of all countries with territorial disputes with other countries."[76] The United States had been against the idea of proceeding with such regional plans, arguing that they might "restrict flexible use of both frequencies and orbital slots,"[77] but progress eventually was made in developing a plan for the Western Hemisphere, as will be discussed in Chapter 7.

The United States, a participant at the 1971 conference, which promulgated Regulation 428A, and the 1977 conference, which reaffirmed and applied it, has maintained that the provision is merely technical and in no way converts prior consent into international law. Stephen Doyle, U.S. delegate to the COPUOS working group, commented: "The ITU requirements for coordination of frequencies should not be taken to imply agreement to the principle of prior consent to international broadcasting."[78] The argument became heated in the mid-1970s when other countries, notably Canada and Sweden, insisted that Regulation 428A had the effect of international law with respect to DBS. In opposition, the United States representative argued that it was necessary to distinguish between direct television broadcasts by satellite that were especially meant for foreign states and those occurring as a result of unintentional spillover; prior consent (were it ever to be adopted) would apply to the former, while Regulation 428A applied to "technical matters," presumably meaning unintentional spillover.[79] In later COPUOS debate, Doyle elaborated on this view, saying that Regulation 428A, taken together with a number of other ITU regulations affecting space-satellite broadcasting, applied only to radio-frequency coordination and notification procedures:

> The prior coordination contemplated and the agreement referred to in paragraph 428A of Article 7 and in Resolution Spa. 2–3 relate exclusively to questions of potential radio interference problems and radio frequency coordination. There is no other dimension or aspect of the agreements contemplated in the language or the history of the ITU Radio Regulations.[80]

Nandasiri Jasentuliyana, a close observer of the controversy over the past fifteen years, gave a different account of how the 1971 ITU conference had arrived at the formulation of Regulation 428A:

> The genuine fear, whether justified or not, of the potential danger of unrestricted satellite broadcasting was responsible for the insistent demand of countries—mainly the developing countries—for the revision of Article 7 of Radio Regulations to ensure that before the establishment of a broadcasting satellite service, there should be a previous agreement among countries who may not want such transmissions or whose service may be disturbed by spillover from such transmissions.[81]

Jasentuliyana regarded the adoption of Regulation 428A as "a turning point in the policy formulations of ITU."[82]

The United States position, that nothing in ITU regulations justifies the acceptance of the concept of prior consent, seems to rest on an analogy with the longstanding ITU regulations of terrestrial broadcasting. These regulations, it will be recalled, prohibit a state from broadcasting to another state, intentionally or accidentally, with a signal that is strong enough to interfere with the domestic broadcasting of the recipient state. If such interference occurs, the offending state is called on to make adjustments. If Regulation 428A is interpreted in the same way, it would mean simply that a broadcasting country would have to select a frequency that is not in use by the domestic broadcasting system of the receiving country. In developed countries, where domestic television broadcasting by satellite is now common, this reading of Regulation 428A could limit any unwanted satellite television penetration, at least with the present level of the technology. But in countries where television is hardly existent, this interpretation of Regulation 428A would in no way inhibit the broadcasting country, since the receiving countries' channels would be unlikely to be in full use by domestic broadcasting, and thus there would be no problem of interference. Another possible motive for the U.S. insistence on differentiating between broadcast signals intended for internal consumption and broadcast signals beamed to selected foreign states is to establish a framework for continued conciliation with Western European countries' interests and for continued confrontation with Eastern European countries. In any case, until 1981 and in spite of urging by the United States, ITU allocations had not yet been made for direct broadcasting across national borders, as was reported to the United States Congress:

> There are no frequency bands available in the radio spectrum for international broadcasting by satellites. Attempts to gain international support for such bands

> in the high frequency, ultra high frequency, and microwave portions of the spectrum for sound broadcasting failed at both the 1977 and 1979 WARC conferences of the ITU, although the United States was in support of this.[83]

The fact that the ITU held back in authorizing allocations for DBS in the HF, VHF, UHF and microwave bands—which are the segments best suited for direct television broadcasting into individual receivers—suggests that the member states of the ITU were exercising an "ITU weapon," withholding legitimation of international DBS until the United States is forced to agree to a prior-consent principle.

Often included with the arguments against prior consent are enumerations of the alternative ways in which nations can prevent the penetration of unwanted satellite signals: for example, jamming; controlling the sale of receiving sets by allowing only the sale of sets that are not compatible with outside broadcasting signals; or monitoring rooftops to prevent utilization of the fairly visible "dish" receivers that have been necessary to receive DBS signals. However, as we shall see, most of these possible forms of control are being weakened by technological advances, making the resolution of the issue of prior consent ever more urgent to its proponents.

Carl Q. Christol, a professor and frequent consultant to the U.S. government, commented on Regulation 428A at the IISL colloquium in 1978, giving what may be presumed to be a restatement of the State Department position. He emphasized that as a technical regulatory agency, the ITU is properly concerned with efficient and equitable use of the two limited natural resources essential for DBS: the geostationary satellite orbit and the radiofrequency spectrum. The proper arena for arguments about "unacceptable ideas and information" had been and should continue to be COPUOS.[84] Christol called attention to the important difference in procedure between these two agencies: in the ITU, issues are submitted to a vote, and the majority prevails; in COPUOS, the longstanding practice has been consensus.

Christol reviewed Sweden and Canada's argument that Regulation 428A gives prior consent the status of international law and took note of the ITU report in 1977 on how it had implemented Regulation 428A in one of its regional plans. He acknowledged that the ITU was, in effect, providing "an equality weapon" to those states unwilling to receive broadcasts.[85] However, such developments did not resolve the problem, he argued, because of the ITU's narrow concern with technical issues and because it lacks "the power to make and enforce decisions relating to national conduct." Christol proposed that a more relevant guarantee was to be found in the Outer Space Treaty's provision that space is "free for use by all states without discrimination of any kind," thus allowing access to radio frequencies in accordance with "relevant prescriptions of international law," which in Christol's view do not include Regulation 428A, useful as it has been. With this convoluted argument, Christol seems to have been suggesting that the Outer Space Treaty protects the United States from any discrimination against its "free-flow" approach.

Wilson P. Dizard updated the U.S. position on DBS in 1980.[86] As a one-time member of the State Department, his views are significant indicators of official policy. In sum, the arguments he gave against prior consent are:

1. It would set "an unfortunate precedent not only for satellites but also in the wider field of world information flow."
2. There is insufficient experience with broadcasting by satellites to justify such action.
3. It could inhibit the advancement and testing of the technology.
4. "Existing ITU regulations provide sufficient safeguards against unauthorized broadcast satellite transmissions," and the ITU technical regulations have "overtaken the need for political principles."
5. It would undermine "international free flow of information and the free exchange of ideas."
6. It would be "unworkable in practice," since a politically incompatible nation could withhold consent thus denying broadcasting to a whole region.[87]

This examination of the problem of interpreting ITU Regulation 428A has provided an opportunity to observe the lines of argument and the motivations of United States spokespersons as they continued their efforts to cope with the DBS controversy. The United States has had to face other difficulties, to which we now turn.

The Geostationary Orbit

In the mid-1970s a curious complication arose that for a time seemed to overshadow the ongoing argument over prior consent and even to unite the United States and the Soviet Union in opposition to the special demands of a group of Third World equatorial countries. The issue had to do with the geostationary orbit (GSO), an important track in space that is the most effective location for communications satellites. The geostationary orbit lies some 22,300 miles (38,500 kilometers) above the equator. Satellites launched into this orbit and traveling at a speed that matches the rotation of the earth appear to remain in a stationary position over the earth. Theoretically, three such satellites set equidistantly in the geostationary orbit over the earth would be capable of broadcasting instantaneously to most of the world's population. The first satellite to be placed in this orbit was NASA's *Syncom 2,* in 1963,[88] and by July 1977, the total was 100. Some experts have estimated that no more than 180 (each separated by 2 degrees) can be placed in the geostationary orbit. Others have predicted that advances in technology will make this limitation irrelevant, partly because the carrying capacity of single satellites will be vastly enlarged.

In part because it has the advanced technology necessary for space launchings, the United States has made the most use of the geostationary orbit. It is estimated that 90 percent of the nonmilitary communications satellites now in the GSO were launched and are controlled by the United States.

Even before the Third World challenge in the mid-1970s, a legal question had been raised by France's suggestion that use of the geostationary orbit could be regarded as an appropriation of outer space, an act prohibited by the Outer Space Treaty.[89] In response, the United States argued that the *use* of an orbit is legitimate under Article II of the treaty and could not be considered a national appropriation.[90] The issue had also been touched on by the ITU conference in 1971, when it adopted provisions to ensure that the placement of satellites in the geostationary orbit (using the time-honored first-come, first-served rule) was not to be considered permanent but could be adjusted to ensure equitable access to all states, a stated ITU goal.[91]

The more serious question about the geostationary orbit arose in 1976, when eight equatorial countries—led by Colombia and including Ecuador, Panama, and Brazil—issued the Bogotá Declaration, in which they asserted sovereignty over the geostationary orbit.[92] They argued that the orbit is not part of outer space, since its existence depends "exclusively on its relation to gravitational phenomena caused by the Earth," and declared that the positioning of satellites over their territories would require their prior authorization. They were motivated in part by the fear that all the orbit slots would be filled by the advanced countries before their own needs could be met. They also regarded the orbit as a "natural resource" that belonged to them (mindful, perhaps, of the sixteenth-century European claims of ownership of the gold of El Dorado, or the more recent assertions by OPEC countries of control of their oil resources).

The demand of the equatorial countries was not accepted at the 1977 ITU conference; as a result, they appended a statement to the final protocol saying that they reserved the right to take "whatever step they may deem fair to preserve and secure the observance of their rights."[93]

The issue raised legal concern and resulted in many restatements of the Outer Space Treaty's authorization of the "free use" of outer space. Not surprisingly, the industrialized countries, including the United States and the Soviet Union, closed ranks on the question, agreeing with the ITU point of view that a plan that assigns orbital positions does not constitute an appropriation, as the equatorial countries were asserting. As to whether the geostationary orbit is or is not part of outer space, this raised the long-dormant question of defining just where air space ends and outer space begins. "Juridically pointless," wrote the Polish space jurist Andrzej Gorbiel.[94] He asserted that the Outer Space Treaty's fundamental reason for existing is to provide international legal principles for the exploration and use of outer space, the whole of outer space, and for states to decide which parts of outer space they would treat as outer space is inconsistent "with the essence of international legal order."

The West German jurist D. Goedhuis saw the Bogotá Declaration as "unsubstantiated" and as a threat to "the principle of freedom of outer space." While agreeing that there is "no rule of positive international law" delimiting air space and outer space, Goedhuis argued that the consensus of states is against state sovereignty being extended into satellite orbit.[95] The general counsel of NASA joined the discussion, stating that the Bogotá Declaration had "no scientific and technical basis."[96] He added that the GSO is but one of an infinite number of orbits; that its uniqueness is attributable to the total gravitational field of the earth and the rotation of the earth; that there is no relationship between the GSO and the territory beneath it, since entry into GSO can be accomplished by launching a satellite from any place on earth; and that the height required for GSO (22,300 miles) is far beyond the altitudes of most satellites. Because of these reasons and because "the legal status of the GSO is well established both by treaty and customary international law," the NASA lawyer opposed not only the Bogotá Declaration, but also any effort to officially invalidate it.[97]

In rejecting the arguments of the Bogotá Declaration, space lawyers recognized the need for broadened and quickened cooperation between the rich and the poor nations. They also set up a committee to once again consider the question of air-space sovereignty.[98]

The significance of the GSO debate for DBS is twofold. First, it underlines the conflict between the underdeveloped nations and the wealthy nations, particularly the United States, which is pictured by the Third World as aggressively expanding its space communications activities and neglecting the enormous communications needs of the less advanced, even impoverished nations. In a more subtle way, it worked to the advantage of the United States by lending the legitimacy of free use of space to the concept of "free flow of information." Second, the convolutions of the legal commentators on the question are new reminders of the ambiguity of international law, which can become a weapon in the hands of those who have the power to proceed with their own interpretation of it.

Radio Broadcasting by Satellite Versus Television Broadcasting by Satellite

It is important to recall that the debates over satellite broadcasting have not always been clearly focused only on television broadcasting. During the 1960s, the rising wave of concern about the possibility of commercial, cultural, or political invasion from the sky seemed to embrace both radio and television signals. The French delegation to an ITU conference in 1963 urged, unsuccessfully, that both television and sound broadcasting from a space object be prohibited. However, the nations and peoples of the world had long been accustomed to a large volume of international radio broadcasting, primarily by shortwave, regulated only minimally by ITU registration and frequency requirements. It was the perceived persuasive power of television as well as the broadcast satellite's unique capability for relaying line-of-sight microwaves that brought the issue to a head. Still, the earliest formal considerations of the matter made use of the ITU definition "DBS"—direct broadcasting by satellite—and that abbreviation has continued in use as a shortcut way of alluding to the activity without specifically indicating whether it refers to television only or to both television and radio signals.

During the 1968 session of the United Nations General Assembly, when the topic of DBS was first discussed, there was general agreement that the issue embraced both radio and television signals.[99] In 1970, when draft principles were first formulated, they called for the regulation of radio *and* television broadcasting via satellite.[100] In 1972, the Soviet Union formally placed the issue before the General Assembly, requesting that the assembly assign to COPUOS the task of developing principles and guidelines for regulating *television* broadcasting via satellite. Thus in the United Nations and COPUOS arenas of jurisdiction, regulation of radio signals appears to have been dropped from consideration at that time. The ITU, however, has never appeared to differentiate between the two forms of broadcasting. Both radio and television signals from satellite are telecommunications; both utilize the VHF, UHF, and microwave frequencies; and both have been deemed to be subject to ITU jurisdiction, in line with its responsibility to regulate international telecommunications.

Deliberations and final actions in UNESCO in October 1972 also dealt with both radio and television signals. It is fairly evident that the U.S. opposition throughout the long DBS debate has been based on the view that none of the discussions is to be construed as in any way applicable to radio broadcasting by satellite, which in the U.S. view is subject to only the longstanding shortwave limitations set by technical provisions of the ITU.

There is much evidence to show that the United States has long been apprehensive about the possibility that prior-consent principles under consideration in COPUOS might be extended to include radio as well as television signals. It is even possible that there is fear that any such guiding principles might be extended to the burgeoning network of public and private telecommunications by satellite; this network encircles the world and feeds information back and forth between banks and businesses and police networks, to mention but a few. Eilene Galloway, former chief staff person for the Senate Committee on Aeronautics and Space Science and a prominent authority on space law, seemed to be making that suggestion in a 1975 commentary on the matter. She speculated that the United States opposes prior consent for "DTB" (one of the few uses of this abbreviation in the literature on this subject) "because this is only one form of space communication technology." Pointing to COMSAT and Intelsat, she added: "This is so strong a combination of governmental and private, national and international endeavor that, I do not believe, they would let *one particular technological use* bring all this into imbalance [italics added]."[101]

Referring, in 1981, to the desirability of seeking ITU authorization for the use of DBS for international radio broadcasting, a State Department official told a House Subcommittee that such a formal request might "result in efforts by other countries to include DBS sound broadcasting in a prior consent regime."[102] William C. Salmon, senior adviser to the undersecretary of state for security assistance, science, and technology, summarized the official U.S. position on the DBS issue before the subcommittee. He reiterated U.S. support of "non-binding principles," provided they did not include "prior consent" and did not apply "to 'other' [audio] broadcasting via satellite."[103]

The Canadian jurists Chapman and Warren had guessed at this motive in 1979. Citing a statement by the U.S. representative in the COPUOS deliberations that the U.S. government "had no plans" to engage in television broadcasting by satellite to another nation without its consent, Chapman and Warren asked:

> If, as it now seems clear, DBS will not be used for transmitting programs in the same unregulated way as international short-wave broadcasting in the High Frequency band (HF) why does the USA continue to oppose any UN principle embodying the right of the receiving state to "agree"? One must conclude that the USA fears that any such principle would be an undesirable precedent that could be applied increasingly to other means of communications or dissemination of information.[104]

In 1981, the U.S. government made public its interest in exploring the use of DBS to supplement the shortwave radio broadcasting that it conducts through the Voice of America, Radio Free Europe, Radio Liberty, and the Armed Forces Radio and Television Service (the Defense Department's network for members of the military services).[105] Salmon posed the following questions as indicators of the issues that had to be studied if direct radio broadcasting by satellite was to be undertaken:

> What are the specific DBS requirements—both sound and television—and target areas, and who are the potential U.S. operators of such systems? Can existing facilities be reduced or eliminated upon implementation of DBS? Are DBS systems truly feasible at a reasonable cost, and if not now, when? What are the technical advantages and disadvantages related to the various frequency bands appropriate for DBS? How susceptible are DBS transmissions to jamming or other forms of radio interference? Will the existing ITU provisions accommodate our future requirements, or will substantial changes and additions be necessary? How will DBS operations impact on other radio services? What are the likely reactions of the world community? What are the possible ramifications in other forums, such as the UN General Assembly and the Outer Space Committee?[106]

The chief reason that Salmon gave for turning to DBS was the need to overcome the reported jamming of American shortwave broadcasts by the Soviet Union. He also cited the congestion of the shortwave frequency bands and the unwillingness of the 1979 ITU conference to accept U.S. proposals for more allocations, including the proposal that the bands reserved for terrestrial shortwave broadcasting be opened up for broadcasting by satellite as well. The implications of Salmon's comments were that, lacking ITU permission, the United States was refraining from the use of DBS for its international broadcasting transmissions, although Salmon testified that it had been made clear to the 1979 ITU conferees that the United States "reserved the right to take the necessary steps to satisfy our requirements in the short-wave band." Possible steps he mentioned included the "use of out-of-band frequencies . . . on a non-interference basis," methods for more efficient use of HF bands, and the use of satellites.[107]

At the same congressional hearing in 1981, James B. Conkling, then director of the Voice of America, emphasized that direct broadcasting by satellite would be "a very major step in eliminating that vulnerability" that the Voice of America confronted by virtue of having transmitter stations distributed around the world, including "a large percentage" in the Third World countries.[108]

The ambiguous legal status of international broadcasting by shortwave has been taken up a few times during this long debate, the experts occasionally expressing puzzlement, occasionally taking it for granted. At the 1973 IISL colloquium, a Swiss authority commented: "Surprisingly, there is still now no consultation or review by any international board concerning the foreign program broadcasting by a state."[109] Christian Patermann of West Germany has pressed the idea that the past fifty years of international broadcasting by shortwave without "any kind of thought of giving permission to another country" have established an "internationally accepted, legally binding practice."[110]

The delegates to the Columbia University forum on DBS in 1973 also touched on the question. Speakers took note of the fact that 99 percent of international broadcasting was by shortwave and that it was conducted by most major governments in order to reach peoples of other countries. The Voice of America, for example, broadcasts in many languages to the Soviet Union, Eastern Europe, China, and Cuba, "the closed societies."[111] Frank Shakespeare, former director of the Voice of America, explained the selection of Voice of America target areas: "Shortwave has no real purpose or effect in . . . the developed and free societies such as those of Western Europe or Japan." Later in the discussion, he added, "The logic of the situation dictates that government propaganda broadcasts serve no function when peoples have a free media and believe it; they serve a very great function when peoples have no free media and therefore are hungry for news."[112] Ownership of radio sets (most with shortwave receiving capability) is widespread in both the Soviet Union (122 million sets in 1975) and China (150 million in 1978). Both governments' domestic radio networks rely heavily on shortwave because of the vast size of the countries and the centralized control of the media. Radios are also in popular use in the Middle East (14 million sets in 1978), in Africa (17 million), and in India (19.6 million).[113] Shortwave radios are not common in the United States, nor are they widely advertised.

As for the fact that international broadcasting by shortwave had come into being under ITU procedures without any discussion of prior consent, Shakespeare speculated in 1973 that "the law never would have been adopted" if the deliberators (presumably meaning the union's member states) had known in the 1920s "how powerful and how all-inclusive radio was going to be."[114] Shakespeare foresaw that the freedom that had characterized shortwave radio was now to be extended to television, thus posing the question: "Will governments permit people in their homes to see foreign pictures in addition to foreign voices? The foreign countries such as Canada and England that voted against us in the United Nations are drawing a distinction. They, in essence, say: 'We believe in freedom, but freedom only for voice. We are too scared of pictures. They are too powerful.'"[115]

Shakespeare's perspective on the difference between traditional shortwave radio and television DBS left out the question of cost. Because of the relatively low cost of shortwave broadcasting, most nations can afford to have their own broadcast facilities. But few have the economic or technological base for launching the rockets that put satellites into the GSO. Another difference is the much higher quality of the satellite-transmitted signal as contrasted with the ground-station-transmitted signal. And the pictures of satellite-transmitted television, compared with the voice of shortwave, not only have greater impact, but also overcome language barriers. Luiz Felipe de Seixas Corrêa of Brazil has commented: "In view of the stage we have reached in the development of international law, it is inconceivable that a field so pregnant with far-reaching implications should be allowed to evolve without the benefit of rules and norms incorporating the views, interests, concerns and aspirations of all states."[116]

Corrêa attributed some of the United Nations' preoccupation with DBS to recognition of "the international tensions which have surrounded short-wave radio broadcasting." The goal is to avoid a repetition of those tensions. "It would be tragic to have the kind of political propaganda one hears on shortwave on home television sets."[117]

The Consensus Procedure

Another area for negotiation which has taken shape as the United States continues to oppose restrictions on the use of DBS is the question of the voting procedures of COPUOS and its subcommittees.

The consensus procedure was adopted in COPUOS very early as a result of diplomatic negotiations aimed at overcoming Soviet opposition to the make-up of the committee, the distribution of committee offices, and the voting procedure.[119] The Soviet Union was insisting that the committee act only by unanimous vote. The United States was opposed. A compromise was agreed on in late 1961: COPUOS and its subcommittees would work toward agreement without voting; but if voting became necessary, the rules of the General Assembly, allowing for majority rule, would apply.[119]

As the debate over prior consent stretched on and on, member states and their jurists began to reexamine the usefulness of consensus, which appeared to be serving mainly the interests of the United States in preventing the resolution of the DBS issue. Not surprisingly, the U.S. space-law experts began to find great merit in the consensus procedure. Two space lawyers wrote that the rule of unanimity effected by consensus has proved to be the best basis for achieving lasting international agreement.[120] They cited the meetings of COPUOS committees on DBS as evidence of the "full extent to which the differing views have been aired, the only consensus being that practically everyone disagrees and will continue to do so."[121]

Eilene Galloway has described the consensus procedure as productive; it resulted in the drafting of four space treaties, all of which have been ratified and become part of international law.[122] She recognized that consensus had not permitted agreement on DBS, on the definition or delimitation of outer space, or on the related issue of the equatorial countries' claim to sovereignty over the geostationary orbit. She attributed these failures not to weakness in the consensus method, but to fundamental differences on substance. In any event, she stressed the wisdom of continuing to follow the consensus procedure.[123]

Carl Christol observed that although the ITU acts by majority vote, COPUOS uses consensus procedures, which he believed to be more suitable to the consideration of abstract social and legal values (presumably including prior consent).[124]

By 1979, the idea was being circulated that the continuing COPUOS impasse—which included not only the unresolved issue of DBS, but also the lack of agreement on legal principles to regulate the remote sensing of the earth's resources from space satellite—might have to be broken by using a majority voting rule within COPUOS and its subcommittees.[125]

In his updating of the U.S. position on DBS, Wilson Dizard noted that the consensus principle was important because it kept debate on the DBS issue "within the relatively restrained precincts of the [COPUOS] Committee."[126] He recognized that a breach might occur and the issue be put before the United Nations General Assembly, where prior consent "would probably be overwhelmingly approved."[127] The United States opposed such a course.[128]

It is evident that the United States, in spite of its preeminent world status, has faced many difficulties in negotiating compromises in the United Nations. Still, until 1982, it managed to stave off any United Nations action to limit the use of space satellites for transmitting television signals into countries that do not want to receive them. In general, its position was perceived by most U.S. analysts as a matter of freedom versus "non-freedom," despite accumulating evidence that demonstrated a far greater complexity of factors. The inability or unwillingness of most analysts to uncover and assess some of the complexity may be, in part, attributed to their method of study as well as to their unrecognized assumptions. The next chapter examines another study perspective.

6

Marxist Social Theory, Mass Communications, and DBS

Some Methodological Considerations

The literature on mass communications has expanded markedly in recent decades, in part as a result of the growing world influence of the United Nations, where Third World members have pressed claims for a new understanding of the meaning and uses of mass communication, and to which the socialist countries, led by the Soviet Union, have brought their arguments about radio "wars," cultural imperialism, and the centrality of state sovereignty. The concern of UNESCO and the United Nations with a "new world information order" has engendered much new, non-Western thinking. At the same time, in the West there has appeared a small but active group of scholars who are thinking critically about the subject of mass communications. A large part of the new literature on mass communications has a Marxist or neo-Marxist orientation.

Much can be learned from this rather diverse body of thought, beginning with the contributions of Marx and Engels themselves and including the many Marxists who have built on their work as well as many others who have borrowed Marxist concepts and used them creatively to develop progressive and nontraditional ways of understanding mass media, especially in its international applications.

This chapter first reviews some important methodological guidelines for investigating the question of transnational broadcasting and the controversy over the direct broadcast satellite, and then proposes a theoretical framework within which the substantive issues of the controversy may be understood.

The Social Context

One essential precept of the Marxist method is that particular phenomena—for example, mass communications—must be seen as rooted in the social and economic processes of the society in which they occur. Communication is "a socially determined phenomenon. . . . One cannot cognize a certain . . . aspect of society deeply without knowing what kind of laws determine social development in general."[1]

Herbert I. Schiller, speaking of the "daily programming" of the American people, regards it as a process "embedded in the unquestioned but fundamental socio-economic arrangements that first determine, and then are reinforced by, property ownership, division of labor, sex roles, the organization of production, and the distribution of income."[2]

Here is Marx's formulation:

> In order to study the connection between intellectual and material production it is above all essential to conceive the latter in its determined historical form and not as a general category. For example, there corresponds to the capitalist mode of production a type of intellectual production quite different from that which corresponded to the medieval mode of production. Unless material production itself is understood in its specific historical form, it is impossible to grasp the characteristics of the intellectual production which corresponds to it or the reciprocal action between the two.[3]

The British writer Raymond Williams believes that studies of major modern communication systems necessarily interlock "historically and theoretically with wider analysis of capitalist society, capitalist economy and the neo-capitalist state." Many such studies "require analysis in the context of modern imperialism and neo-colonialism, to which they are crucially relevant."[4] A Polish scholar, Walery Pisarek, has touched on this issue, pointing out that to treat mass media, or individual newspapers or broadcasting stations, as isolated autonomous phenomena is to neglect key connections—for example, that "the social role of the same newspaper or even the same article changes according to the changing social conditions in which the readers live."[5]

This methodological insistence on the interrelation between the phenomenon and its social context has broad implications. To begin with, any examination of international broadcasting by satellite must take into account the relationship of broadcasting to the twentieth-century socioeconomic structure of advanced capitalism and to the social relations that have evolved within that structure. It must also consider how socioeconomic structures and social relations are organized in states that are perceived as having systems different from or hostile to capitalism. Finally, it must acknowledge the varieties of social contexts within which world populations interact with one another.

Another application of this methodological insistence occurs at the level of the researcher herself, who must confront her relation to the class structure in which she functions. Her research is necessarily "characterized by class orientation both at the stage of conceptualization and of formulation of conclusions." It is not possible to "be an impassive and detached reporter of social and political phenomena in general [or of] the process of communication in particular."[6] Put another way: "An individual has a certain location within the material world and the social relations concretely operating in it."[7]

The Importance of History

Another tenet of the Marxist method is that issues can be understood fully only by tracing out their historical context; that is, by identifying the linkages between the phenomena under examination and the socioeconomic trends and events that led to them. An example is suggested by Armand Mattelart's appeal to researchers to pay more attention to connecting

> what are known as "content" analyses and analyses of production of discourse with the changes taking place in the regime of capitalist accumulation, with the changes in the relations of the dominated classes with the bourgeois State, with the changes in the composition of the ruling classes and their systems of alliances and with the evolution of current technical advances.[8]

By applying the historical method to an analysis of the role of the United States in international cable and radio communication (Chapter 2), it is possible to discern and highlight certain trends that would be lost or covered over in ahistorical, here-and-now analyses. For example, we may recall the long struggle between state and private enterprises over control of communications technology and processes and the eventual resolution of the struggle with the apparent freedom of the industry from government interference. But under examination, there was found close cooperation between, if not integration of, industry and government. Moreover, by relating broader historical trends to evolving communication politics, we are able to see significant parallels between economic and political expansionism and the rise of dominion in the field of telecommunications.

Another example is found in the work of J. D. Bernal, who looked at the history of science not as isolated human endeavor, but as events deeply embedded in socioeconomic processes. In his outline of how scientific discovery is transformed into technical application, Bernal emphasized the determinant role played by the prevailing economic structure—in Marxist terms, the "mode of production." "Science is thought of and paid for in proportion to its contribution to increased value of products or reduction of costs."[9] Bernal was concerned with many of the issues raised in this study, especially with those discussed in Chapter 2. He highlighted the debilitating effects of monopoly, the erosion of the original purposes of patents when they are turned into controls that thwart rather than promote invention, and the disinterest in using scientific inventions to improve the lives of people above and beyond the profit-making interest. Bernal's work is significant in the history of science because of his insistent intertwining of the emergence of science and invention with the concurrent social attitudes and values.

History as Process

Implied by Mattelart's and Bernal's stress on the importance of history is another fundamental methodological tenet of the Marxist approach—that the human condition is best comprehended as a continuing unfolding. This emphasis on process embraces the concept of dialectical contradiction, which views phenomena as constantly being transformed, constantly in tension or conflict. Marx used and developed this Hegelian idea in his various social analyses, as exemplified in his familiar formulation:

> At a certain stage of their development, the material productive forces of society come in conflict with the existing relations of production, or—what is but a legal expression for the same thing—with the property relations within which they had been at work hitherto. From forms of development of the productive forces these relations turn into fetters. Then begins an epoch of social revolution.[10]

A present-day commentator on this paragraph from Marx explained:

> The Marxist view is that the tension between productive forces and productive relations . . . leads to class conflict. This class conflict occurs within (1) the productive relations, (2) all social and political institutions, and (3) the realm of ideas. Eventually the class conflict causes revolutionary change in productive relations, institutions, and ideas. The *new* productive relations, *new* institutions, and *new* ideas resulting from the revolution lead to a new expansion of the productive forces (including technology), although the form of the new is still influenced by the old forces.[11]

An application of this approach is found in the hypothesis that "the communications revolution is shaped by regional and class struggles, by powerful capitalist forces molding that revolution to meet accumulation and legitimacy needs, and non-capitalist forces resisting hegemony and using information resources to build a new social order."[12]

This study has pointed out the incomplete character of the DBS controversy, emphasizing the ongoing diplomatic negotiating processes; the evolving nature of international space law; the adaptiveness, even inconstancy, of sociological communication theory; and the ever-shifting play of international forces.

Critique as Method

Another characteristic of the Marxist method of social analysis is the effort to go beyond appearances, to tease out meanings, to be exhaustively critical of the givens and of the assumptions and findings of contemporary social science and social theory. In their search for latent, or unspoken, meanings, Marxist students of communications have uncovered, among other things, the widespread and uncritical acceptance of market terminology in American communications sociology. Communication itself "is conceived in the American communication research . . . as an *exchange of* information, meanings, etc., between people." The choice of the market analogy is symptomatic of something deeper: "It is precisely exchange that gives the visible ethos to a capitalist society in general."[13] But "when communication is reduced to the scheme of an exchange act between two persons it becomes cut off from those basic forces whose manifestation it is, and consequently its true understanding is made impossible." Thus we become aware of the misleading character of the words that we are obliged to use. To equate the words *communications* and *exchange* is to equate the power of the sender and the receiver; the term *media* is used to lump together global news-gathering corporations with the village gazette; the "human right" to enjoy "free flow of information" is a right enjoyed by only the powerful news conglomerates or nation-states, but is made to appear as an individual's right to exchange information with other individuals. In the whole construction, the "receiver" as free agent in a free exchange process takes on a grotesque unreality. Mattelart commented:

> The term "means of communication" is rarely used by bourgeois sociologists, and has been hidden behind the economically and politically neutral terms "communication media," "mass communication media," and particularly "mass media." Thus the materiality of the means of communication is obfuscated, and the immaterial aspects are emphasized; the final result is that the product of the media is substituted for the production process of communication.[14]

The insistence on a critical stance with regard to both the terminology and the substance of mainstream social science also permits a reexamination of the manifold claims of empiricism. While frequently useful and illuminating, empiricism cannot be substituted for critical thinking and reflection, nor can it be permitted to obscure the web of relationships that tie a phenomenon to its social, political, and economic context. Veikko Pietilä, a Finnish authority on communications research, described Western empirical sociology's approach to communications research: "Mass communication and communication in general are seen . . . as a certain class of social phenomena, discernible as such from social phenomena of other kinds. . . . Therefore the fact that communication is always a manifestation of something else becomes obscure and easily ignored." He provided an example: "Because the wielding of power usually requires communication, the phenomenal approach easily conceives the power relations as communication relations, which they on no account are."[15]

Mattelart proposed that the empiricism of mainstream communication sociology has the effect of precluding any notion of class. Bourgeois empiricism, he said, puts forth the following position:

> Reality is just as it appears, the meaning of the message is limited to its explicit content, all relationships established between individuals and social groups are transparent, and the exchange effected between the producer and the consumer, between the transmitter and the receiver, between the center and the periphery is an exchange carried out under conditions of equality and objectivity.[16]

This harmonious scenario is undermined by Marxist social analysis, which forces recognition of class differences based on power and property disparities. Mattelart continued, "The notion of class contradicts the sacred truth according to which, since everyone is free to receive information, everyone is also free to express and transmit his or her own information to others."[17] In this critical framework, the existing worldwide inequalities of access to telecommunications facilities—from sophisticated broadcasting satellites to community telephone lines—can no longer be concealed by or counterbalanced with empirical data.

Validation Through Praxis

Marxist-oriented researchers do not agree on all the components of the Marxist method. One of the tenets on which there is some divergence in opinion is the requirement that what is theorized must be validated through a praxis that is aimed at transforming society, "Social practice is the only criterion of veracity."[18] Or, speaking of those who study mass communications: "The degree of our understanding [of the nature of the political struggle within communication] has a profound effect on how we further elaborate our communication practice and fight that imposed by the dominant classes."[19] Also on this point, Herbert Schiller, who sees the domain of communications research as "nothing less than the physical, structural, institutional bases of a system of domination that operates through its impact on human consciousness," calls for "critical opposition to the foundations of what is now hailed as 'the information society.'"[20]

Summary

Taken together, the components of the Marxist method provide a useful beginning for understanding the present controversy among nation-states over the possibility of unregulated expansion of television broadcasting across national borders. They make clear that the issue cannot be disconnected from the capitalist forms of television broadcasting, ownership, and control which exist in the United States, the most powerful and influential participant in the controversy. Nor can it be separated from the economic and social-class structures that prevail in the United States and in the countries to which direct television broadcasting by satellite would be beamed. Furthermore, the direct broadcast satellite cannot be perceived solely as a new and ahistorical technology. While it represents an enormous advance in radio capability, the issues that it raises can be discerned as early as a century ago, when nation-states began to use the electromagnetic telegraph to facilitate the necessary exchange processes of industry and to strengthen the arm of the state. A knowledge of the tensions and rivalries among nations and among corporations that arose concurrently with the development of the telegraph, the laying of the undersea cable, the emergence of wireless, and the rise of international radiotelephone and shortwave provides a basis for an analysis of the forces at work in the DBS controversy.

In the examination of international space law (Chapter 4), it was noted that legal authorities from the Soviet Union argued that international law is itself a form of struggle. This interpretation is foreign to Western legal theory but is provocative and illuminating if reexamined through a lens that sees history as a process that involves interacting class forces as well as nation-states and that, when the strife reaches a certain level, takes the form of sweeping social change. This view, that social existence is to be understood as class forces in conflict, is useful as a guide to the comprehension of radio and television broadcasting as we know them. True, non-Marxists as well as Marxists have commented repeatedly on the unrealized but potentially liberating quality of the electronic miracle of the radio signal, pointing out how it can give voice to those long silenced, empower the powerless, and bring education and improved hygiene to the illiterate and impoverished. But in the Marxist framework, this "potential" is regarded as transformative, even revolutionary, and as proceeding from the ongoing material development of production processes. How does this happen? A sketchy, and therefore oversimplified, answer is that mass communication radio and television, as presently organized under capitalism, must necessarily reach ever-larger audiences in order to realize ever-larger profits or achieve ever-more efficient cost–benefit ratios. But the world mass population to which radio and television are directed may thereby gain media literacy and, with it, a degree of social awareness and self-consciousness that could undermine existing power arrangements, including the control and use of radio and television. A contemporary example[21] might be the current movements in South Africa and in Central America, with their tumultuous mix of new consciousness, liberation assertions, and increasing sophistication engendered by exposure to mass media, which some resisters now use as a counter weapon.

The critical stance that Marxism requires of any analysis of capitalist society is useful for unmasking the ideological commitments that protect and overarch much of contemporary sociological theorizing about mass communications. Examples of these uncritical commitments were offered in Chapter 3.

Finally, the methodological insistence on social transformation implies acceptance of the fact that research always has a political purpose, implied or overt. This relationship between scholarship and politics has been illustrated in the United States with the proliferation of government- and military-sponsored scholarly research since the end of World War II. Another example is this study, which, while designed to trace the history and development of mass communication across national borders and to clarify the substance of the debate over international control of such broadcasting, bears also the writer's purpose of addressing the concerns of the millions of people who are increasingly subjected to external economic and communications interests that seek to control, direct, and profit from their existence and to rename, coopt, or suppress the opposition they might offer. Speaking for communications scholars who identify with the critical approach, one among them recently wrote:

> By attempting to trace the historical nature of communication, information and culture, our need is eminently contemporary and political: to learn from this

> experience, to better distinguish between the "organic" and "conjunctural" movement in communication (to use the words of Gramsci), to better discern the specific nature of each class's communication, and especially to draw the political implications flowing from present tendencies in communication.[22]

A Theoretical Framework

As we just saw, it is difficult to separate the methodological components of Marxism from its theory. We have touched on much that is theoretical, and now turn to a fuller discussion of a number of core concepts of Marxist social theory that seem especially significant to the understanding of mass communications development. Separating them for purposes of discussion should not be allowed to obscure their interrelation within a social whole.

Fundamental to Marxist social theory is the understanding that the mode of production is a key determinant of the way a society thinks about itself. The mode of production is the way in which a society meets the material needs of its members; it includes the *means* (tools, processed materials, physical structures) by which the production of goods and services is accomplished and the *relations* between the participants in the production process (some as owners; most as nonowners, who sell their labor to the owners). In Marx's words:

> In the social production of their life, men enter into definite relations that are indispensable and independent of their will, relations of production which correspond to a definite stage of development of their material productive forces. The sum total of these relations of production constitutes the economic structure of society, the real foundation, on which rises a legal and political superstructure and to which correspond definite forms of social consciousness. The mode of production of material life conditions the social, political and intellectual life process in general. It is not the consciousness of men that determines their being, but, on the contrary, their social being that determines their consciousness.[23]

To illustrate, we might say that the capitalist mode of production, which has reached a high point of development in the late-twentieth-century United States, fundamentally shapes our daily lives, political and social relationships, and beliefs, as well as the institutions by which we more or less maintain social order. Peter Berger and Thomas Luckman, both of whom are not Marxist, explain that what concerned Marx was that "human thought is founded in human activity ('labor' in the widest sense of the word) and in social relations brought about by this activity."[24] Another formulation: "Ideas do not arise in a vacuum, but within a particular social environment on a particular economic base."[25]

Looking at contemporary mass communications in the United States in this light, we recognize their primary role as an integral part of a capitalist economic order: they are a form of production whose purpose is to increase the capital invested in them, in order to further expand the production cycle, to make possible new investment, and to enhance the wealth of the owners of the capital. "The cultural-communications sector of the world [market] system necessarily develops in accordance with and facilitates the aims and objectives of the general system. . . . Cultural-informational outputs are largely, though not entirely, determined by the same market imperatives that govern the overall system's production of goods and services."[26]

A closer look at how the agencies of mass communication function leads us to see a second purpose, in addition to economic production. Because their commodities are ideas, the media owners not only sell them for profit,[27] but also use them, or arrange for them to be used, to reinforce the sets of beliefs that are necessary to the ongoing operation of the prevailing mode of production. This double value of the output of the media industry may be understood more clearly within a theory of society that posits both a material base of production and a culture of social values and institutions that in various ways arises from or reflects the conditions and needs of the material base.

Culture and the Material Base

In the Marxist world view, capitalist production (the dominant economic form of life in the twentieth century) is understood as the material base of society; the culture, or sets of beliefs and social institutions, that it in large part engenders or undergirds is perceived as arising out of, constructed on and interacting with the material base. Turning to mass communications as they have developed in the United States, it may be seen that they actively function in both spheres of this model—in "the base," as commodities produced for consumption, and in the culture, or "superstructure," as carriers of ideology.[28] This duality appears to differentiate the mass media industry from most other forms of material production. An analogy can be made with architecture, which not only is architectural design, but also "constitutes by itself a very efficient means of propaganda, and as such . . . has been used by Roman emperors, French kings or Latin American dictators. . . . Architecture is a form of 'mass media,' and as such is both infrastructure and superstructure, industry and art, technology and ideology."[29]

To perceive the influential and determining role of the capitalist mode of production as it works itself out in the field of electronic and mass communications is to better understand the conditions we are examining. Transnational industry's compelling need to expand, and the dual nature of that expansion when it occurs in the mass communications industry (ideas as commodities and ideas as legitimation), is revealed as driving the communications industry ever deeper into not only world markets, but also the social and cultural spheres of the lives of the peoples who make up those markets. The communications analyst Cees Hamelink provides this assessment: "The massive one-way traffic of socio-cultural perspectives manufactured in the metropolis structures the symbolic environment of Third World populations." The market needs of transnational firms mandate "an optimal levelling of cultural differences in order that national characteristics do not jeopardize the unity of the transnational system." The advertising and marketing branch of the international information industry may be seen as an important carrier of this process as it "propagates worldwide the concept of the global shopping center."[30]

Seen in this perspective, the significant growth and concentration patterns of the media industries (75 percent of the information market is controlled by some eighty transnational corporations; 10 to 15 percent of the world's largest corporations are involved in the international information trade) must be understood not as a plot to undermine foreign cultures and economies, but as the logical outcome of a particular mode of production. Many useful studies of the transnational media industries, lacking the larger framework of critical social theory, fail to take account of the interaction between the needs and development of a production system and the effect of that system in the cultural and ideological realms.[31]

Although the "determining" impact of the expanding capitalist mode of production on value systems and institutions of social order is profound and fundamental, it is not by any means a one-way relationship. Precisely the kind of relationship it is has become an increasingly contentious point of debate among Marxists. Some question the degree to which the "determining" process is actually effective or "necessary" or "inevitable": grave misgivings are expressed about the cause-and-effect rigidities of the more orthodox, or classical, Marxists; Soviet theorists often appear to bear the brunt of these critiques. In general, the new critics tend to be drawn ever farther from the fundamental concerns of Marxism: class conflict, exploitation, resistance, and social transformation.[32] On the positive side, however, their work sometimes has led to a wider recognition of the complexity and rich suggestiveness of the relationship between material forces and consciousness.[33] The British Marxist Ray Williams is one of those who has enriched the sometimes too reductionist model of a "determining base" and a "determined superstructure," or ideology. Drawn to Gramsci's theory of hegemony, as a more comprehensive and subtle conceptualization of how social-control processes saturate "the whole process of living," Williams calls for methods of analysis that will reveal not only the hegemonic limits imposed by capitalism, but also the significant openings and breaks that occur, in many forms, in the struggle to create an alternative hegemony.[34]

While Marxist theory accords to the material base of society the dominant and determining role over the ideas and ideology of society, a fuller unfolding of Marxist sociology shows that ideas and ideology can influence material conditions. According to a contemporary formulation, "Human activity forms the conditions under which human ideas come into being, but once these ideas exist, they constitute an important force (for example, the ideas of liberty, equality and fraternity)."[35] Marx's thinking about the distinction between the two spheres may be seen in his description of how social revolution occurs: not only in the "economic sphere" of production, but also in "the legal, political, religious, aesthetic or philosophic—in short, ideological forms in which men become conscious of this conflict and fight it out."[36] We will, in a later section, return to this possibility of conflict and to the forms that it may take.

The State

The Marxist perspective that differentiates the material base of society from the cultural and ideological superstructure is helpful in illuminating the dual character of mass communications as well as the imperatives of capitalism to create world markets and the consumerist mentalities that make those markets possible. It also provides a framework for considering the nature of the state. To oversimplify Marx's original analysis of capitalist society, it may be said that he viewed the state as located in the sphere of the superstructural, identifying it, in effect, as an organized form of social control that is installed and maintained by the dominant forces of a specific time and society. In the years since Marx so described it, the state has been much theorized about, both by non-Marxist liberals seeking to shore up the long-prevailing view of the state as neutral arbiter among a plurality of forces and by Western or neo-Marxists. The latter, taking into consideration the events of contemporary history, including the emergence of socialist states, have sought to determine the degree to which the state might be independent of the dominant productive forces. A number of subtleties and refinements have emerged in the basic theory, but there is still wide agreement among Marxists that the modern capitalist state is, in fundamental ways, an apparatus shaped and controlled by the ruling elite, although, as in Marx's original view, tension and even struggle arise in the course of exercising and retaining that control.

The relationship between the state and the dominant mode of production is an important one for understanding contemporary mass communications and, especially, the role of the United States in its long debate over the direct broadcast satellite. As we have seen, mainstream social scientists and observers have tended to identify the opposition of the United States to prior consent with its support of the abstract concepts of freedom and the rule of constitutional law, accepting these as ahistorical principles without recognizing their connection to a material base. Thus they leave unexamined such questions as how "free flow" is invoked to serve the market needs of the communications industry, how the portrayal of "free" is sanctified by that industry, or, perhaps most important, how that industry effectively influences and shapes the official government policies that it requires to ensure its market success.

A recent example of the media industry's influence on government policy may be seen in the withdrawal of the United States from UNESCO. A major factor in bringing about the withdrawal was the strong opposition of newspaper publishers and news-gathering and broadcasting corporations to what they perceived as UNESCO's move in the direction of inhibiting their news businesses, evidenced by the call for a new world information and communication order (NWICO) to redress imbalances in access to information and in the system of news collection and distribution. The media industry's opposition was formalized in what it called the Talloires Declaration. Issued in 1981, the declaration states that "denying freedom of the press denies all freedom of the individual" (paragraph 4) and stresses "the importance of advertising as a consumer service and in providing financial support for a strong and self-sustaining press," without which "the press cannot be independent" (paragraph 7).[37] It continues a long tradition. From the advent of transnational electromagnetic communication, through the establishment of the Radio Corporation of America, the rise of the Federal Communications Commission, the development of international and domestic broadcasting regulation, and the emergence of huge private telecommunications conglomerates, the decisions made by the communications industry soon turn up as decisions of the U.S. government.

Nowhere has the fusion of state and capital been more manifest than in the creation of COMSAT, a profit-making semiprivate corporation created by public law and assisted by vast outlays of public tax money. The triumph of COMSAT is exemplified in Intelsat, the multinational entity that it created in the 1960s and that is now put forward as a model for entrepreneurial cooperation in transborder telecommunications. A document from COMSAT and Intelsat enthusiasts proclaims the superiority of "the competitive market philosophy" in comparison with "the natural monopoly philosophy" or "common heritage of mankind" approach, to which some states are attracted as they seek to control their own telecommunications via space satellite. The benefits of Intelsat were enumerated: its "investment shares, earnings, and to some extent voting power correspond directly to the percentage of use a nation makes of the Intelsat system. Thus, while all mankind can benefit from the Intelsat system, there is no overemphasis on guaranteeing absolute equal treatment of all participants." This remarkable statement of the perceived dangers of equal treatment was part of the advice given to delegates who were to represent the United States at another round of ITU negotiations on the GSO.[38]

The symbiosis between the state and the holders of economic power may be observed in how telecommunications are used by the United States as the legitimate guardian of U.S. interests at home and abroad. This use requires that it have control of military communications networks on a world scale, as, indeed, the Pentagon does—both independent of and in conjunction with the corporate telecommunications networks. The recognition of the state as adjunctive to the imperatives of capitalism makes it possible to understand motives and events that are usually ignored. For example, the drive for nuclear military strength sufficient to ensure U.S. superiority in the world is a necessary condition for the unrestrained movement of capital to wherever it can make a profit, including the flow of mass communications production via satellite. While the governmental uses of radio for international broadcasting are different from the commercial uses—fewer claims to "news neutrality," financial support through taxes rather than through advertising, less subtle persuasion techniques—their alignment with the dominant economic forces is no less clear. As a component of the state, the Voice of America uses the mass broadcasting technology to diffuse and make acceptable an ideology that supports the dominant mode of production in the United States. The more militant voices of Radio Free Europe, Radio Liberty, Radio Martí, and soon-to-be Radio Free Afghanistan[39] and Radio Maccabee (directed toward Jews living in the Soviet Union) serve to intensify the Voice of America messages in broadcasts aimed at socialist countries, with the purpose of undermining and destabilizing their social cohesion and governmental structures. The military and intelligence uses of mass communications across national borders provide the most aggressive wing of this psychological warfare, again with the purpose of preserving and protecting the needs of capitalism for markets and raw materials. Miguel Blandón, director of Radio Sandino and Nicaragua's representative at a recent meeting of the Nonaligned Movement's Radio Broadcasting Organizations, provided this comment (duly reported by the CIA's research service):

> One of the most important subjects, as we see it, is the use of radio broadcasting as a means of aggression against our countries. Nicaragua, being a weak country, does not have much technology and therefore cannot defend itself in the broadcasting field, and it is the victim of flagrant attacks from the United States, a very rich world power that can afford the luxury of installing transmitters in neighboring countries along our border in order to penetrate and try to influence the most backward sectors of the population. Therefore, this is, for us, one important and hot issue, because it is very much linked with the military attacks that we are suffering, especially from the northern border. As we all know, these attacks are always linked with counterrevolutionary radio broadcasts that prepare the conditions so that these people may operate more easily.[40]

Hegemony and Class Formation

The Marxist idea that the mode of production shapes consciousness is not to be understood as a fixed or abstract or one-way relationship, but as a relationship that arises anew out of a play of forces under specific conditions and in specific historical periods. The influence of material conditions manifests itself in varied ways among groups of people holding different social positions and reflects their interaction with one another. It is in such a context that we now take up the Marxist concept of class.

The complexity of Marx's concept of class and class formation is well illustrated by this frequently cited passage from *The Eighteenth Brumaire of Louis Bonaparte,* in which he describes the small-landholding French peasants of the 1850s:

> The small-holding peasants form a vast mass, the members of which live in similar conditions but without entering into manifold relations with one another. Their mode of production isolates them from one another instead of bringing them into mutual intercourse. The isolation is increased by France's bad means of communication and by the poverty of the peasants. Their field of production,

> the small holding, admits of no division of labour in its cultivation, no application of science and, therefore, no diversity of development, no variety of talent, no wealth of social relationships.[41]

The result is that the "great mass of the French nation is formed by simple addition of homologous magnitudes, much as potatoes in a sack form a sack of potatoes." He concluded:

> In so far as millions of families live under economic conditions of existence that separate their mode of life, their interests and their culture from those of the other classes, and put them in hostile opposition to the latter, they form a class. In so far as there is merely a local interconnection among these small-holding peasants, and the identity of their interests begets no community, no national bond and no political organization among them, they do not form a class.[42]

From his observations in these passages and elsewhere in his work, it is clear that Marx saw class formation and class consciousness as flowing from a complex social process. The fact of being a worker, whether on a small holding or in a factory, does not by itself confer consciousness of class status or a sense of solidarity with one's fellow workers as against the dominant class. Consciousness and its potential as a foundation for class action is realized only through the knowledge gained by working with others, communicating with a broad sector of fellow workers, developing political and national awareness, and organizing. Marx's theory builds directly on the importance of material conditions—the kind of work one does and the level of subsistence one has—as primary determinants of one's ideas about oneself: "The mode of production in material life determines the social, political and intellectual life processes in general."[43] But as modern life has made very clear, factors other than material conditions influence the intellectual processes necessary to the development of class consciousness. Central among these are the ever-increasing spread of literacy and the availability of information. With the growing democratization of Western society, a process begun only two centuries ago by the French Revolution, people gained access to information that had been reserved for an educated elite. Newspapers, broadsides, and pamphlets became commonplace.[44] Concurrent with the growth of capitalism on a world scale, humankind has gained even greater access to information, via a mass media that includes not only print, but also the visual and voice transmissions made possible through radio waves. As we have seen, a special characteristic of the new mass media—"the consciousness industry,"[45] as the contemporary press, radio, and television are called by one media critic—is that it is owned primarily by members of the dominant class and that its purpose is to maximize profits and advance ideas that support the capitalist-economic system. Or, as Marx and Engels put it: "The ideas of the ruling class are in every epoch the ruling ideas: i.e., the class which is the ruling material force of society is at the same time its ruling intellectual force."[46]

Gramsci, who had been a journalist before his long imprisonment, was among the early-twentieth-century Marxist theoreticians who took up the question of how class consciousness was to be achieved among the masses. His early life in Italy, where he had felt the full force of the dominant class as it prevailed over the movement toward socialism, led him to a profound rethinking of how authority is exercised in the ideological arena of a society. Gramsci delineated two sections of that arena: "one that can be called 'civil society,' that is the ensemble of organisms commonly called 'private,' and that of 'political society' or 'the State.'" In the former ("civil society"), the dominant group holds power through indirect means. In the latter ("political society"), it holds power through "'direct domination' or command, exercised through the State and 'juridical' government."[47] His elaboration of the forms of indirect control is significant:

> The "spontaneous" consent given by the great masses of the population to the general direction imposed on social life by the dominant fundamental group . . . is "historically" caused by the prestige (and consequent confidence) which the dominant group enjoys because of its position and function in the world of production.[48]

Gramsci's formulation permits us to consider how the prestige of the mass media industry confers legitimacy on both its message and its message senders. That this appears to occur in a "civil" or "indirect" or seemingly "private" realm, beyond the command of government, gives it added persuasiveness.

Pietilä suggested that as the bourgeoisie confronted the rise of socialism in the twentieth century, it sought to counter it by relying on ideological control, exercised predominantly through the mass media: "This ideological regulation, connected with the fetishized nature of the apparent social reality, accounts for the fact that it is not easy for the working class in capitalist societies to avoid adopting bourgeois ways of thinking."[49]

Successive Marxist and progressive theoreticians, as they have examined mass communications, have taken up the Gramscian propositions about how the dominant forces exercise hegemony. Miliband approvingly cited G. A. William's definition of the process as "an order in which a certain way of thought and life is dominant, in which one concept of reality is diffused throughout society . . . informing with its spirit . . . all social relations, particularly in their moral and intellectual connotation."[50] In the West, this diffusion of certain ways of thinking is generally accomplished without overt directives. It is unusual to find as clear a statement as the following policy issued by Procter and Gamble to the networks with which it advertises: "There will be no material [in programs sponsored by Procter and Gamble] that may give offence, either directly or by inference, to any commercial organization of any sort . . . [or] which could in any way further the concept of business as cold, ruthless and lacking all sentiment or spiritual motivation."[51]

A more usual approach is an admixture of the manifest with the latent message, as was described by Leonardo Acosta:

> This technique manipulates associations by projecting images indirectly. The overt message generally reflects the official ideology of the system (liberty, democracy, individualism), while the latent injects the spectator with the mimetic image of the system and the submission to its paternalistic and authoritarian values. The latent message of many television dramas and comedies is essentially the "uselessness of rebellion" and the "convenience of conforming."[52]

Gaye Tuchman offered this formulation: "American television perpetuates hegemony; it adumbrates political and intellectual discourse. It not only buries dissent; it buries the possibility that new ideas may emerge."[53] Put another way: "The consequence of the hegemonic process is that the majority of the population is largely unaware of alternative values and alternative readings of history."[54] "In the absence of visible alternatives, no mass-based opposition emerges and the structure of control is able to continue unchallenged."[55]

Stuart Hall described the functioning of the mass media in the cultural sphere as being

> crucial to the reproduction of the dominant culture and ideology that it conveys. This is done by providing the social knowledge through which the reality of others is perceived, so that an imaginary coherence is built, in which the contradictory elements are held within the frames of the dominant culture and ideologies. This transforms them into a formidable apparatus for the building of consensus and legitimacy.[56]

Put more briefly: "The power to define is the power to control"[57] or more poetically, from Samuel Beckett: "I am walled around with their vociferations, none will ever know what I am, none will ever hear me say it. I won't say it. I can't say it. I have no language but theirs."[58]

An example of the hegemonic process at work is found in the phrases "free flow" and "freedom of information," the key words carefully chosen to support the U.S. position in the DBS controversy. Drawing on the connotations of "free speech" and its long constitutional history in this country, the terms enjoy an almost independent existence, an "absolute, static, indivisible reality"[59] that obscures the source of the "flow," the beneficiary of the "freedom," and the contents and effects of both. "Free flow" is advanced as a pure "good," reified for use as a weapon to stave off reasoned discussion. Once invoked, it suspends all alternatives; even to suggest one is to move outside the allowable bounds of discourse. Early use of the term appears to have stemmed from the postwar efforts of the media industry to reestablish its world standing. By 1950, "free flow" had successfully been installed as a fundamental tenet of UNESCO's work on broadened flows of information. Only in the 1970s was the term challenged and elaborated to include a recognition of the demands for greater balance, as described elsewhere in this study.[60]

In accomplishing or deepening the hegemonic sway of the dominant group in a society, few new technologies have proved more effective than the electronic broadcast media. Their fetish status and unique quality of "speaking" on behalf of those who control the broadcast system is described in this striking Third World perspective:

> Radio enters the home or accompanies the listener anywhere as a form of property, ever smaller thanks to the transistor. It is, moreover, an object that talks, that transforms its bearer into a permanent listener. The philosophy of consumption is thus fused with that of the mass media and takes the form of a fetish object which, dressed in the latest style, is presented and sold as the finished product of a system; a product which in turn, speaks in defense of that system. As could be expected, the culmination of the mass media is television, which synthesizes the main qualities of the film and radio (sound, image, color, words, music). The television set has become the new ritual magic center of the home.[61]

The fetish metaphor was much elaborated by Marx in his analysis of the capitalist mode of production. He used it effectively to discover underlying assumptions, long hidden by revered myths. Succeeding generations of progressive thinkers have identified the "fetishizing" techniques used in various capitalist activities, including mass communications. Armand Mattelart, speaking of communications in Chile in 1971, noted:

> Every fetish reflects an entire rational corpus of mechanisms whose purpose is to obscure the social relations of production prevailing in a given society. [In capitalist society] the fetish of communications conceals the repressive and manipulative power of the dominant technology of dissemination (a veritable new *productive force*) [Mattelart's italics] and characterizes it as a force for liberation and happiness. It is in this guise that it is presented to the dominated.[62]

Mattelart described what he called the "fetishization" process, which occurs as "living people are metamorphosed into 'things' (factors in production), and things are given life. Thus money 'works,' capital 'produces.' In the same way, the media 'act.' "[63]

Turning to the direct broadcast satellite and its use on a global scale, it is evident that such mass communications as are likely to be carried by direct broadcast satellite from the United States will reflect established mass media patterns and will thus become extensions and reinforcements of the hegemonic control now exercised within the United States—adjusted, of course, for non-North American audiences. The organization and ownership of the giant broadcast and communications corporations based in the United States and the imbalance in the flow of their transmissions (newspapers, radio, and television) across international borders have been the focus of a number of studies, including those of Herbert Schiller,[64] Tapio Varis,[65] Alan Wells,[66] Richard Bunce,[67] Jeremy Tunstall,[68] Anthony Smith,[69] and S. B. O. Gutto.[70] Their findings make it clear that today's transnational conglomerates have spread a worldwide net of consumer advertising, film, and radio and television broadcasting that, with subtle hegemonic suasion, compels conformity without consciousness. It is an operation that has not gone smoothly, as has been shown in earlier chapters of this study. Subsistence conditions among vast populations of the world have sometimes provided barren ground for the spread of Western ideology; the ideology itself has sometimes been wrongly interpreted, taking the shape of new demands; the very process of modernizing has often engendered new or alternative ideologies. Nor can it be ignored that during this century, one-third of the world's population has come to be living in countries that are formulating social and economic systems that differ from and are competitive with those of the West. Whether these developments represent class formation and class consciousness as emergent from material and social production processes and, if so, whether they can be subverted or redirected by persuasion and the calculated distribution of benefits, or even by force, are questions that are yet to be answered; the control of world information networks, including the direct broadcast satellite, is one of the essential factors affecting the outcome.

Class Conflict and Resistance

Class conflict has been evident throughout history but is more apparent and fully developed during the dominance of the capitalist mode of production when the difference between classes is a requisite condition for production. Simply stated, the owners of the means of production constitute a class, and those who work but do not own the means of production constitute another. This division into two classes and the antagonism between them may be observed at critical points in history, as dominated or oppressed groups become aware of their conditions of existence and in various ways seek to transform them.

Simple as it may appear, this formulation gains depth and complexity when it is understood from a theoretical standpoint that holds that what appear to be separate entities are actually interrelated and interdependent: each defines the other, and to such a degree that they form a kind of conflictive unity. Hegel's famous formulation of this dialectic in terms of the master–slave relationship was to provide a basis for the later elaboration and use of the dialectic by Marx and Marxists as a way of understanding social development.[71]

Without venturing into the philosophical and psychological controversies that still envelop the Hegelian contradiction, it will suffice to point to the usefulness of the notion of how the slave and master, in their unending search for mutual recognition, engage each other in struggle to undermine or negate their accepted roles and in the process achieve different conditions of existence, including self-understanding and self-determination. The idea that domination sows the seeds of its own negation or that servitude breeds self-awareness and the conditions for its own transformation has won wide acknowledgement and has many times in history proved its validity.

In this study of how and whether global controls are to be imposed on new forms of mass communication and persuasion, the imperatives arising out of conflict and resistance are discussed in a number of dimensions. Consider, for example, the relationship between the United States and the Soviet Union. The hostility against the Soviet Union that is expressed by the leaders of the United States and is fostered successfully (largely through the mass media) among the people of the United States is generally justified in terms of such issues as the history of Stalinism, the absence of formal democracy, the discrimination against Soviet Jews, the ongoing state prohibitions against subversion and anti-Sovietism within the Soviet Union, the alleged pro-revolutionary proselytizing conducted by the Soviet Union in other parts of the world, and even the allegation that the Russian Revolution has been betrayed. Seldom mentioned as a basis for opposition to the Soviet Union is the mode of production that it has adopted, which functions without the necessary components of private capitalist enterprise and in direct contradiction to the capitalist ideology. Whether or not the Soviet Union is succeeding as a "workers' state" (the arguments would take us too far from the immediate issue), the simple fact of its existence may be seen as providing the underside, a noncapitalist side, of the present class conflict. When developed by Marx and others in the nineteenth century, class struggle was generally perceived as occurring not between nations, but between classes—the industrial capitalists versus the industrialized workers, regardless of nationality. However, the lens has widened in the past century, bringing into focus more and more of the world's peoples and their systems of production. Accordingly, it is possible and evocative to perceive the present "superpower" struggle between the United States and the Soviet Union as fundamentally a conflict between a capitalist-dominated economy and a noncapitalist-dominated economy. This is not to minimize all the other differences between the two nations, including different levels of ideological control and of economic and political development. But it is necessary to bring out this core difference, so widely ignored in contemporary analyses. In a perspective that recognizes United States hostility toward the Soviet Union as being rooted in class struggle, the intensity of the radio and television war becomes more understandable. At stake are fundamental forms of social and economic organization, the very modes of production, each with its reflected values, benefits-distribution systems, and method of social control.

From another angle, the conflict between classes may be seen to underlie the relationship between the United States and the Third World, where steady increases in the amount of available information and in awareness lead to eruptions of guerrilla warfare and other forms of rebellion. Marx's critique of capitalism, itself drawing on the work of many earlier writers, Ricardo and Smith among them, stimulated many generations of radical analysts, from Lenin, Rosa Luxemburg, and Rudolf Hilferding to Paul A. Baran, Samir Amin, Immanuel Wallerstein, Andre Gunder Frank, and many others both in the West and in the institutes and universities of the Soviet Union and other socialist countries. Drawing on Marx's outline of the preconditions necessary for capitalist production—raw materials, a system of private property, "free" workers willing to sell their labor, and markets—these later theorists deepened Marx's critique by adapting it to observable changes in advanced capitalist production—for example, the intensifying concentration of industry, the technological advances in productive methods, the enormous expansion of markets, the transnational division of labor, and, the state, a powerful institutional component, that provides the force as needed to maintain and control markets and raw materials.

Within the school of Western Marxism, there has developed in recent decades a perspective that sees late-twentieth-century capitalism as a world system, made up of "core or advanced capitalist" societies, "periphery or pre-capitalist agricultural/extractive" societies, and the "semi-periphery intermediary societies."[72] According to this perspective, the core societies extract surplus value from the periphery, thus keeping these societies in a state of underdevelopment. The semiperiphery societies also yield surplus value to the core and buffer it from the periphery states. Whether this trend toward a world market with the core societies as the chief beneficiaries is regarded as imperialism or, less pejoratively, as the fuller development of the wealth of nations, it is clearly the pattern that modern capitalism must follow in order to ensure itself a continuing increase in return on investment, without which capitalism cannot survive. Fredric Jameson sees the present period as constituting "the purest form of capital yet to have emerged, a prodigious expansion of capital into hitherto uncommodified areas," including even the penetration and colonization of the unconscious via the rise of media and advertising.[73] The United States, as the most powerful core society, has become the protector of capitalist expansion, an activity that it carries out by insisting on the market model as the *sine qua non* for any developing country that wishes to avoid military advisers, CIA penetration, radio jamming, loan defaults, currency destabilization, or other forms of intervention. At the same time, it has led the drive to incorporate less-developed societies into the capitalist market structure, including the accelerating use of Third World workers for factory and field production. It is the often unrecorded resistance to these processes that makes up the underside, the other half, of the whole. When international mass communication is studied within this framework of domination and resistance as a unity, the DBS option is revealed in its full political dimensions.

Vincent Mosco and Andrew Herman have examined aspects of this issue. They cite three communication infrastructures—the submarine cable, the earth-orbiting satellite, and the computer—as primary examples of core-society technology used to achieve and shore up imperialist expansion. When linked to more traditional mass media (advertising, newspapers, radio, television), these infrastructures become "not merely the generator of a vast accumulation process, but also instruments of world system legitimation."[74] In 1965, speaking to a conference of the American Legion, David Sarnoff of RCA provided an astonishingly frank summary of the role of communications: "Supremacy in the domain of communications, which is one of the keys to the domination of space, translates today into political, military, economic and social supremacy over all the nations of the world."[75] His son and successor, Robert, was more restrained, predicting that "information will become a basic commodity equivalent to energy in the world economy" and "will function as a form of currency in world trade, convertible into goods and services everywhere."[76]

In considering some late-twentieth-century manifestations of class conflict, we have proposed that it may be perceived in the East–West confrontation between two profoundly different economic and social systems and in the intensifying conflicts between the United States and the nonaligned nations. Conflict and contradiction may also be discerned as arising within the cultural sphere, where the weapons are the words and technology of mass communications themselves and where the boundaries of nation-states are less significant. Here Gramsci's formulations are particularly relevant, since he identified the different forms of resistance that confront the hegemonic imposition of consent. While the working class (the "subaltern group") is constantly being subordinated "to the active hegemony of the directive and dominant group," Gramsci emphasized how its own efforts at autonomy are constantly being "reborn in other forms, as parties, trade unions, cultural associations."[77] In mainstream studies of mass communications, these possibilities and manifestations have received small attention.

Bertolt Brecht was among the first to identify the resistance potential of radio. His 1927 "Proposals to a Radio Director" demanded the airing of parliamentary sessions, of courtroom trials, of unrehearsed interviews and debates. In 1932, he outlined his "Théorie de la Radio": "Non seulement que le public soit éduqué, mais qu'il soit lui-même éducateur."[78] He identified one of the fundamental contradictions of the new apparatus of radio—that its owners needed people like him to replenish their stock in trade, the very stuff that was to be broadcast over the airwaves: "Mais ce n'est pas du tout notre tâche de renouveler par des innovations sur la base de l'ordre social existant ces institutions idéologiques. Nous devons plutôt par nos innovations ébranler leur base: pour les innovations, contre le renouvellement!"[79]

The use of the electronic media in the cause of popular resistance, even revolution, was described by Frantz Fanon in his account of the Algerians' war against the French.[80] Fanon was aware of the interrelation of the oppressor and the oppressed and sought to show "that the challenging of the very principle of foreign domination brings about essential mutations in the consciousness of the colonized, in the manner in which he perceives the colonizer, in his human status in the world."[81] It was that change in consciousness that led to the Algerians' ever-wider use of the clandestine source of news, opposing "the enemy news with his own news."[82] Fanon continued: "In making of the radio a primary means of resisting the increasingly overwhelming psychological and military pressures of the occupant, Algerian society made an autonomous decision to embrace the new technique and thus tune itself in on the new signaling systems brought into being by the revolution."[83]

The Algerian experience has been repeated a number of times as the decolonizing of Africa and Latin America has accelerated.[84] At the same time, the immense electronic technologies of the dominant states have been further developed in order to catch up to and capture clandestine initiatives. In his study of international broadcasting, Julian Hale offered a report and analysis of the turbulent world of clandestine radio, which he estimated in 1975 was broadcasting about 5 percent of the output "in the worldwide communications battle."[85] One aspect of radio as a subversive instrument that has not been realized, except in *Dick Tracy* and specialized mobile units, is its potential as transmitter as well as receiver. Hans Enzensberger pointed out that "every transistor radio is, by the nature of its construction, at the same time a potential transmitter; it can interact with other receivers by circuit reversal," a development that, he argued, "is consciously prevented for understandable political reasons."[86]

With the gigantic leap in communicative power represented by the satellite and its direct broadcast capability, the industrialized nations gained an edge over their less developed opponents. Schiller speculated that the ongoing class conflict of our time is becoming more intense specifically in the communications-cultural sphere, as the dominating class uses its "enormously expanded and totally penetrative informational apparatus" to maintain its longstanding privilege.[87] Schiller proposed that the advent of space communications has heightened the conflict, pointing to how the new technology has "spotlighted the overall issues of structural control, financial and technological dependency, and the dramatic side of information penetration from the heavens."[88]

In an analysis of the relationship between culture and dependent development, Raquel Salinas and Leena Paldán outlined some possible conditioners of resistance—for example, the level of organizational experience that the working class may acquire as it undergoes industrialization, or the degree of polarization between competing sectors of the population, which solidifies the political alignments of both the dominant "internationalized sector" and the excluded sector. A third possible source of alternative or resistant ideologies is the cultural industry, where those who create the cultural products may "escape the limits and models imposed both by the metropolis and the local dominant culture by transforming their own practices into the expression of the lived reality of the dominated."[89] The suggestion that the person hired to create cultural products for the dominant class may become proletarianized echoes Marx's provocative equation between the teacher in the "learning factory" and the wage laborer in the real factory.[90] Finally, Salinas and Paldán remind us of the new audience toward which the media producers must direct their work, often without cognizance of its "multiple realities" and "multiple histories of accumulated significations whose importance to the interpretation of the media messages cannot be neglected."[91]

In the present struggle over communication policies, Salinas and Paldán, like others, recognize the ambiguous role of the resisting state, which at the same time that it opposes external domination often proves willing to align with new forms of internal domination. Nevertheless, they argue that at the present level of the conflict, the demands of resisting states "contribute to the steadily weakening hegemony of the dominating classes in the cultural-ideological field."[92]

In summary, once the liberative potential of the mass media is examined in the context of dialectical relationships, the present conflict over world-wide information flow and the technology of the direct broadcast satellite takes on a new significance. Along with its enormous economic importance, as a means to speed the movement and widen the markets of capital, the direct broadcast satellite is a new and superior channel for responding to political needs to contain the rise of liberation movements or, conversely, to strengthen the energies of resistance.

When Hegel described his master–slave model, he posited no certain outcome. The dialectical engagement could lead to death, to new forms of subjugation, or to social transformation. A better understanding of the conflict between contemporary world forces may serve to shape our participation in that engagement.

7

Technology Update and Some Concluding Observations

Technological Developments

The technology and manufacture of the instruments of electronic telecommunications have taken on awesome dimensions. Moreover, the meshing of the telecommunications satellite with the power of the computer is resulting in vast new communications capabilities and a constantly growing variety of applications. Motivation for these developments comes primarily from the needs of world capitalist markets and the requirements of military research and procurement, but the technological progress is endorsed by various non-Western leaders and by many others who hope to use telecommunications directly to improve trade and standards of life and culture. This dialectical interaction between technological development and sociohistorical determinants has far-reaching implications for the world of the future.

Before proceeding, it would do well to emphasize again the two major kinds of space-satellite telecommunications services that have been provided for by the radio regulations of the ITU: the Fixed Satellite Service (FSS) and the Broadcast Satellite Service (BSS). As noted earlier, the FSS is satellite communication between fixed points on earth and includes, among other things, telephony, television distribution, and data transmission. Its use has expanded very rapidly in the industrialized countries since the ITU first allocated spectrum space for it.

The Broadcast Satellite Service (BSS) accommodates signals that are broadcast to designated earth stations for redistribution to the general public. One application of BSS is DBS—direct broadcast by satellite—in which, as we know, the signal is intended for direct reception by the general public without an intermediary earth station.

While the emphasis of this study has been on who is to control the use of DBS, it is clear that such issues as prior consent and equitable access to the geostationary orbit have implications for other types of international satellite telecommunications.

As an update, it may be useful to review some of the most recent technological advances (without attempting to provide full technical explanations) and to suggest how they may be affecting U.S. policies and world politics with regard to the development and utilization of communication by satellite, including DBS.

One notable trend is in the direction of ever-smaller, less costly earth-station dish antennas, made possible as knowledge and experience is gained in the utilization of the ultrahigh-frequency range ("Ku-band") now being used for most satellite radio and television transmissions (between 11 and 13 gigahertz, a gigahertz being 1 million waves per second). It is possible to receive signals with a dish less than 2 feet in diameter, and the lower cost and smaller size of antennas will allow an ever-larger portion of the world's population to acquire them.

Ways are also being found to overcome the language barriers that might have affected domestic applications of DBS in a country whose inhabitants speak different languages or that might have inhibited telecasting into foreign countries. Through the combination of analog and digital transmissions (transmitting the video in the analog mode and the audio in the digital mode), the video may be accompanied by audio in several languages, one of which may be selected by the viewer.

Related to the development of the digital approach for satellite television is the emergence of what is called "enhanced," or "high-definition," television (HDTV), a technique that results in a superior television picture, comparable with film in quality. These kinds of video transmissions were not possible prior to the use of the extremely high microwave frequencies (in the gigahertz range). To receive "high-definition" signals, however, requires a different type of television receiving set from those now in use. Economic and marketing forces are building toward worldwide agreement on a single standard for this new technological breakthrough in improved video quality.[1] When standards were set for present-day television, the standards adopted in Western and Eastern Europe called for higher definition than the standards adopted in the United States and Canada, with the result that European television sets cannot receive American signals. If a single worldwide standard is adopted, international direct television broadcast by satellite would not face barriers to reception because of incompatible signals—a significant development.

Significant, too, is the rapid evolution of sound broadcasting via DBS. Segments of the spectrum have been designated by the ITU for experiments in direct radio-signal transmissions by satellite. However, to receive such signals, new radios have to be acquired or frequency converters have to be purchased in order that conventional radio sets may be adapted to the much higher wavelengths. Given the shortage of prime slots in the geostationary orbit, it seems likely that radio DBS—because it is cheaper to put into operation and requires much less precious bandwidth space than does television—will move forward more rapidly than television DBS. Its development must be in accord with ITU rules and regulations, if existing international law is to be observed.

Another rapidly developing aspect of satellite communications is in the area of mobile receivers (for example, briefcase telephones or two-way phones in airplanes, boats, trucks, trains, and other vehicles). Their feasibility has already been widely demonstrated. One authority comments: "The domestic mobile satellite proposals in the United States and Canada expect about 200,000 users in the two countries by the end of the century. They [consumers of the service] would be offered nationwide mobile radio telephone, voice and digital dispatch, interactive data, and perhaps surveillance and other services."[2] While here projected mainly for commercial use, this two-way form of satellite radio communication makes possible an as yet unassessed mode for "talking back" or "talking around" the traditional one-way form of radio broadcasting, not unlike citizens band radio.

Consideration must also be given to the know-how and new tastes that are being acquired as a result of certain unofficial international applications of DBS. Projects at the United States Information Agency, Columbia University, and Vassar and other colleges are now monitoring Soviet spillover television by tracking the *Molniya* satellites by computer and capturing signals meant for domestic Soviet consumption.[3] In an undertaking that must require some tricky adaptations, a young Soviet electronics amateur has done something similar with the USIA's Worldnet program, which is beamed by satellite to the United States Embassy in Moscow. At Vassar College, members of the Russian Department view the Soviet television fare on Japanese-made television sets, which have been manufactured to receive both European television signals, including the Soviet system, and American television signals. The spreading awareness of what is available in the sky is also evidenced by the dish antennas visible on rooftops and in farmyards in North America.[4]

On the negative side, public concern has surfaced over the possibility that concentrations of very large commercial satellite-communication dish antennas, as in Vernon, New Jersey, may be emitting harmful spillover radiation microwaves and causing an abnormally high number of birth defects and miscarriages.[5] It is a reminder of the many unknown dimensions of microwave development.

While there are many indications that direct television and radio broadcasting by satellite is increasingly feasible, other modes of expediting telecommunications are being explored. These include the light-transmitted fiber-optic system, which promises terrestrial, static-free telecommunication on a vast scale and thus might be a less costly successor of or supplement to the communications satellite. Another frontline of research is the work being done in both the United States and the Soviet Union on ionospheric modification. Such modification would make the ionosphere function as a reflector of the ultrahigh-frequency waves that carry television signals, a process that would appear to duplicate the work of space satellites at far lower cost.

Issues in the United States

Closely related to the growing range of technological developments is the political and economic conflict over how the new communications processes are to be distributed and utilized, especially in the implementation of transborder broadcasting. Since the opportunities, arguments, and competition reach beyond national boundaries, it is useful first to consider the new problems confronting the economic and political system of the United States and then to review their unfolding in the larger scene of world politics and the United Nations.

Within the United States, there is continuing commercial interest in the possible utilization of DBS as a domestic enterprise, although its low profitability and high cost, compared with the profitability and cost of existing methods of satellite and terrestrial broadcasting, make it less attractive than it might have seemed when it was new. In addition, the proliferation of cable television has made it more difficult for the establishment networks to deliver large audiences to advertisers, a fact that has put a damper on new expenditures for DBS and has led the networks to expand into the cable business themselves.

Another form of intercorporate competition is taking shape between, on the one hand, those interests (including the promoters of "free flow") that see opportunities for making capital gains by marketing cultural programs and advertising in the Third World via DBS programming emanating from the United States, and, on the other, those interests, such as Hughes Aircraft and Ford Aerospace, that see opportunities for selling huge national telecommunications and DBS systems directly to Third World countries, which would then operate the systems independently (and, presumably, be less vulnerable to penetration). This argument has taken on concrete dimensions in the form of sharp demands by the big manufacturers of telecommunications systems for direct U.S. loans to Third World countries, which would then buy the systems from these manufacturers—a not unprecedented appeal to the state to facilitate the business enterprise.

Conflict between different forms of capital is also evident in the present situation of Intelsat. Although it began under the influence of COMSAT, a quasi-private U.S. corporation, Intelsat now reports to a stockholder membership of 110 nation-states, which jointly own 16 geostationary communications satellites that carry 66 percent of all international telecommunications traffic (and 98 percent of intercontinental and transoceanic television). Under these circumstances, Intelsat has understandably embraced an internationalism that sometimes puts it in opposition to U.S. capitalist expansion. Until recently, Intelsat was the major carrier of U.S. intercontinental telecommunication-network services. Internationally, it has faced competition from only its much smaller counterpart, Intersputnik, which links countries of the socialist world. But during the mid-1980s, Intelsat was successfully targeted by U.S.-based entrepreneurs who wanted to compete with it by offering lower cost service for the ever-growing communications demands of multinational corporations. Opposing this development, the Intelsat leadership cited the original requirement that no Intelsat member should use or set up competing systems without prior consultation on matters of compatibility and avoidance of economic harm to the Intelsat global system. Intelsat also pointed out that loss of its heaviest users might require it to charge higher rates to its remaining customers, precisely the countries least able to pay. The controversy is engaging the energies of Congress and the FCC as well as Intelsat members and leaders. Intelsat representatives were reported to have met in Moscow in March 1985 with Intersputnik representatives, presumably to discuss what would occur should higher Intelsat tariffs cause its members to consider joining Intersputnik.[6]

Intelsat also operates coordinated domestic telecommunications systems for twenty-six of its member states, an operation that is experiencing competition from the trend toward independent domestic (Brazil) and regional (ARABSAT) satellite systems. As noted above, the selling of such large systems is of great importance to satellite and telecommunications conglomerates in the United States.

Another issue for the development of space telecommunications in the United States is the often ignored matter of financing. Like the railroad, trucking, and airplane industries, space exploitation has depended on vast outlays of government financing—the contribution of the taxpaying public. No private-enterprise substitute has been found for the research gains, technological prowess, and almost limitless resources of the Defense Department and NASA. The relationship that this interdependence engenders between the public and the private sectors can be a source of tension. One example may be seen in the contradictions arising over information flow, when national-defense interests are cited by the Defense Department as justification for restricting commercial sales overseas or for restraining international scientific-exchange programs conducted by researchers employed by private corporations or universities.[7] Another point of strain may be seen in the growing public awareness that the huge tax subsidies for space-communication development are serving more to facilitate the accumulation of profit than to accomplish widespread noncommercial low-cost domestic distribution of the benefits of satellite communication.

While the use of DBS for domestic broadcasting is unlikely to occur soon in the United States, its use for foreign broadcasting is attractive to the particular needs of commerce. For example, tobacco companies, with their marketing expertise, have increasingly found that they can successfully advertise their product in Third World countries to make up for the shrinking American market.[8] At the point that DBS is able to deliver consumers more cheaply than can other communication modes (for example, foreign-based advertising agencies using indigenous communication networks; sales of films, television series, and magazines; or news-distribution agencies), it will become irresistible to advertisers and will probably be used as shortwave radio was during the early 1940s, when, with FCC authorization, radio voices from the United States sought to persuade Latin American listeners to buy Camel cigarettes and Adam hats and to stay at the Waldorf Astoria.[9]

The most immediate prospect for both radio and television DBS utilization is in the U.S. government's longstanding program of international broadcasting. In the early 1980s, the USIA received increased funding as part of a projected $1.3 billion improvement program. Expansion of its coverage included resumption of radio broadcasts to Western Europe, presumably to counter the unsympathetic treatment accorded the United States and NATO by the European peace movement.[10] Also under consideration were ways to make the Voice of America's radio message available in the United States, which is now forbidden by law. The possibility of extending USIA's global reach via radio DBS has been under consideration for at least five years; the House of Representatives held hearings on the matter in 1981, as reported in Chapter 5. The House committee report endorsed DBS as likely to provide the Voice of America, Radio Free Europe, and Radio Liberty with bigger audiences, better signal quality, freedom from jamming, and reduced vulnerability for its overseas transmitters. Currently, NASA is conducting a $1.5 million study to determine the ways in which DBS could be used as part of the administration's commitment to upgrade the Voice of America. One continuing problem for radio DBS is the transition from conventional shortwave radios (it is estimated that 1 billion adults in the world have access to shortwave radio) to radios adapted in such a way that they can receive signals in the high frequencies used by space-satellite transmissions.[11] Another problem is the relocation of far-flung overseas personnel—no small matter in the case of the large numbers of expatriate Russians and East Europeans who make up the fiery anti-Communist contingent at the Munich headquarters of Radio Free Europe and Radio Liberty. Yet another problem is, of course, international law and the regulations of the ITU, although these affect radio DBS only to the extent of controlling the segments of the overcrowded broadcasting spectrum that may be used. Until the matter is resolved, the USIA is continuing to negotiate for, acquire, and equip large, heavily guarded terrestrial transmitters (in Costa Rica, Puerto Rico, Belize, Sri Lanka, Morocco, Thailand, Botswana, and Israel, among other places).[12]

The Voice of America's move into direct television broadcasting by satellite is confronting different problems from those raised by the radio DBS transition. The United States Advisory Commission on Public Diplomacy, a body named by President Reagan to oversee and advance the work of the USIA, recommended in its 1985 annual report that the USIA, "as a high priority" and "with the support of other government agencies, plan for the establishment of a worldwide system of television broadcasting to enable U.S. leaders to speak directly to audiences in every region of the world." This to be accomplished with "innovative use of the latest satellite communications technology" and at least "within the next decade."[13] According to the report, a start had been made in the form of the Voice of America and USIA television broadcasts (presumably, the Worldnet program), which had been going through the PTTs of the receiving countries (a procedure that appears to skirt the prior-consent issue). It was expected that the USIA would shortly "place dish antennas on the roofs of our embassies," since direct transmissions would "result in considerable savings."[14]

A 1983 MIT study of the feasibility of using DBS for Voice of America international broadcasting, commissioned by the Voice of America and written by Ithiel de Sola Pool, among others, recommended that the United

States consider development of an international commercial satellite "common carrier," which would rent channels for DBS broadcasts from different countries. The report proposed that millions of listeners would be willing to invest in the new UHF sets that would be required to receive the broadcasts. As for the Voice of America's traditional function: "The American view (via VOA) will come out well in such a forum, as it always has."[15]

The 1985 report of the United States Advisory Commission on Public Diplomacy alluded briefly to legal problems that confront the "goal of a worldwide direct broadcasting service." Among these are the ITU regulations and the prior-consent position of the majority of the members of the United Nations. As we know, the United States does not regard these limitations as applicable to the intended use of DBS for television. However, that Washington policy makers regard the question seriously is evident in the State Department's commissioning of a detailed analysis of the pros and cons of the continued participation of the United States in the ITU.[16] The authors of the report recommended against withdrawal, emphasizing that it would close off the United States from new markets for sales of technologies and from direct participation in training programs of Third World technicians and that it might increase the general risk of "spectrum anarchy." However, the withdrawal of the United States from UNESCO in 1984 and from the jurisdiction of the World Court in 1985 is a strong indication that, in what appears to be a reversal of postwar support for collective international approaches to global problems, the dominant forces in this country are increasingly ready to seek separation from world bodies that do not accept its policies and recommendations.

The military advances in satellite communication have been only briefly discussed. But this update should not neglect the two secret space launches that the Department of Defense conducted in January and in October 1985, both directly involving the expansion of a global "command and control" broadcasting satellite system.[17] The importance of this military global communications and surveillance capability grows in tandem with the dangers of nuclear war and a total breakdown in relations among nations.

Issues in the United Nations and International Agencies

It is to the broader plane of cooperation among nations that we now turn, for a brief look at recent attempts to formulate communications and information policies that can serve a hoped-for peaceful world development. Perhaps the most significant event, although hardly commented on by communications authorities in the United States, occurred on December 10, 1982, when the United Nations General Assembly voted to adopt a resolution approving and accepting guiding principles for the use of direct broadcast satellites for television, thus ending further consideration of the matter in COPUOS (text of document, Appendix C). Among the principles adopted was the prior-consent protection sought by the vast majority of United Nations members. The vote was 107 for, 13 against (including the United States, Japan, Israel, and ten Western European countries), and 13 abstentions. The resolution was sponsored by eighteen African, Asian, and Latin American nations and Rumania. Reporting the event, the *New York Times* followed the traditional United States interpretation by writing that the "General Assembly, endorsing a curb on the free flow of information, today declared, in effect, that all nations had the right to veto any incoming television broadcasts by satellite from abroad."[18] It also reported that one of the resolution's advocates, Ernesto Rodriguez Medina of Colombia, "suggested that major nations opposing the resolution while alleging so-called violations in the free flow of information apparently 'did not desire any bias to their technological development.'" The *Times* reporter also wrote that although General Assembly resolutions are not binding, this one was "widely expected here to serve as a basis for negotiations on an international convention."[19]

In meetings of the fifty-four-member COPUOS since 1982, the possibility of developing a treaty proposal to transform the nonbinding prior-consent resolution into international law has not been pressed by its advocates. Instead, the committee is continuing its long-stalemated debate over the point where the atmosphere ends and outer space begins. The Soviets are proposing 70 miles above the earth as the cutoff point, but the United States opposes any fixed point, arguing that high-tech spacecraft like the space shuttle are making space flight possible at lower altitudes than had been previously thought possible (recalling that aircraft cannot fly much higher than 80,000 feet, which is presumed to be the cutoff point for air space which is, by ICAO treaty, subject to national sovereignty).

COPUOS is also continuing its effort to regulate satellite sensing, an issue that has proved as difficult to resolve as prior consent, with the developing countries unsuccessfully asserting access rights to information about their resources that is gathered by satellite surveillance.[20] For a time, COPUOS members sought to consider the danger of the militarization of space raised by the "Star Wars" initiative, but a threatened walkout by the United States removed that item from the agenda and affirmed an agreement that the only United Nations body that can consider it is the United Nations Conference on Disarmament.

The prior-consent guiding principles for DBS adopted by the General Assembly refer to only the direct broadcasting of television signals intended for individual receiving sets; they do not apply to the Fixed Satellite Service, in which the sending and receiving entities negotiate the terms of the exchange in advance. When Fixed Satellite Service broadcasting affects more than one nation, its planning must be carried out in accordance with the growing body of regulations of the ITU. As this work has proceeded, the focus of action around the issues of transborder broadcasting has tended to shift from COPUOS to the various working groups of the ITU. Under a new director, Richard Butler (an Australian whose selection was opposed by the United States), the ITU, which has virtually the same membership as the United Nations, has appeared to take a more vigorous part in the international politics of telecommunications. It has also been playing an important role as an advocate of the Third World countries' campaign for full observance of applicable international law.

One of the central and immediate concerns of the ITU is the equitable distribution of the slots in the geostationary orbit (GSO), in which more than 150 satellites (mostly belonging to Intelsat or the United States) now operate. The argument is infused with the questions that make up the daily agenda at the United Nations and in confrontations between the United States and Third World countries: how to make the wealth and technology of the West more accessible to these countries without furthering their dependence on and penetration by Western culture and capital or diminishing the opportunity for returns on capital investments. In the case of the GSO, the United States seeks to ward off what it calls "a priori" planning, which would allot orbital slots to countries on an equal basis and without regard to their readiness to use them. The United States argues that advancing technology is allowing closer placement of satellites (2 degrees apart instead of 3 or 4) and thus will make more room in the precious orbit when it is needed. But other nations are pressing their demand for an end to the first-come, first-served policies that have prevailed in the ITU. The matter was debated at a crucial ITU meeting held in Geneva in August 1985 and is to be resolved, if possible, at a continuation of that session set for 1988.

Through programs and seminars, the ITU also is committed to providing the developing and communications-hungry nations with more technical know-how—including information on standardization and on the integration of space communications systems with land networks—and with help in obtaining outside funds for investment in telecommunications infrastructures. This commitment has helped build a counterforce to U.S. policies, which are often based on dazzling technical superiority. One result of the ITU initiative was the 1985 report of the Maitland Commission, proposing the Center for Telecommunications Development within or close to the ITU. Such an independent center has little appeal for the American telecommunications industry, which is seeking to maintain its longstanding dominant position relatively free from international oversight. One of the new telecommunications newsletters comments: "The aim [of U.S. policy makers] is to make it impossible for ITU Director General Richard Butler to gain personal or practical control over Center resources as they come into being."[21]

The strengthened prestige of the ITU will have to be reckoned with by the United States if it decides to proceed in defiance of international opinion on the issue of DBS. The problem is not only the ITU provision known as Regulation 428A (Chapter 5) and the recent General Assembly affirmation of prior consent, but also the long-established ITU jurisdiction over the designation of segments of the radio spectrum for specified purposes. The ITU member states have agreed to designate only very limited segments of the spectrum for DBS, and the designated frequencies are closely restricted to prevent interference with existing terrestrial radio services. The United States is pushing for extended designations, but the ITU conferees are cautious.

And most ITU references to the matter remind readers that when designations are made, they are to be utilized in accordance with the ITU coordination procedures and regulations. As for television via direct broadcast satellite, the ITU in 1979 allocated a segment of the 12 gigahertz band for DBS, most of which is to be shared with Fixed Satellite Service. The ITU holds another trump card—the physical relationship between a frequency designated for broadcasting from outer space and the reception of its signal on earth: the higher the frequency, the less the power needed to generate the kind of signals that can be picked up by the small earth-station antennas used for individual reception. In the long view, there seems to be a tug of war between, on the one hand, U.S. commercial and government interests seeking to keep open and to broaden the possibility of DBS international radio and television broadcasting, and, on the other, the vast majority of ITU member states seeking to regulate that possibility through limits on the spectrum segments that satellite broadcasting must use and on the U.S. dominance of the geostationary orbit.

The ITU member countries of Region 1 (Asia) and Region 3 (Europe and Africa) have moved to resolve the two problems of designating space broadcasting frequencies and assigning GSO slots. At the ITU 1977 meeting, they set aside for Regions 1 and 3 a portion of the spectrum's ultrahigh-frequency waves (11.7 to 12.7 gigahertz) for use by Fixed Satellite Service or Broadcast Satellite Service (DBS). The member states of these two regions delineated 252 "service areas" for reception of satellite broadcast signals. They also assigned slots in the GSO to 150 nations and territories. The United States insisted on postponing the development of a plan for Region 2 (the Western Hemisphere) until 1983. That plan has now been negotiated, with emphasis on interim arrangements that permit development to go forward pending more permanent implementation of fixed spectrum allotments and GSO assignments.[22] What is not resolved by any of these plans is the matter of direct television broadcasting by satellite *between* regions, except that the plan of Region 2 may not contain modifications that would interfere with Broadcast Satellite Service in Regions 1 and 3.

As part of the United Nations' stepped-up interest in redistributing access to space technology and space telecommunications, a multinational space conference, "Unispace '82," was held in Vienna in 1982. Under the sponsorship of the ITU, COPUOS, and UNESCO, a group of eighteen experts (none from the United States) wrote a report on the feasibility of national use of DBS for educational purposes, including the possibility of international or regional DBS systems.[23] It provides essential information to countries or groups of countries on how to proceed and urges that developed countries, international organizations, and United Nations programs provide the needed technology and financing. It also proposes that a cost-effective way for poorer nations to get direct broadcast satellite service would be to form an international, multichannel, multisatellite time-sharing system that could be supervised by one of the existing organizations (ITU would seem to be a possibility) or by a new one, presumably sponsored by the United

Nations. Similar suggestions are emerging for a communication satellite to be owned by the United Nations itself, including the possibility of financing it through a tax on the international telecommunications equipment or on geostationary privilege.[24] Like the Law of the Sea Treaty or the proposed treaty to authorize and regulate mining on the moon, the idea of United Nations ownership of a communications satellite moves the issue ever farther in the direction of international lawmaking and collective regulatory regimes based on equal respect for United Nations member states, a direction in which the leaders of the United States do not wish to or are not able to proceed.

UNESCO, undaunted by the attacks on it from the West, is continuing its longstanding promotion of DBS, including a project for the development of a "small low-cost receiving earth station for eventual endogenous production" to be used for the redistribution of direct satellite broadcasting signals to individual receivers.[25]

Summary

The accelerating promotion of space-based telecommunications services for developing nations and the growing complexity of international regulatory and supervisory procedures are limiting the various options open to the United States. If it wishes to proceed within the emerging international legal framework, it will find it increasingly difficult to justify the penetration of the television airwaves of other countries without their prior agreement. Overwhelming technological, financial, and military superiority in space telecommunications may preserve an unfettered approach for a period of time. But the demand of other countries for more equitable access to the channels and to the substance of information needed for their survival and development is inexorable, and contention and negotiation over the issues raised in this study will continue for a long time. It is a disquieting future, with uncertain outcomes.

By way of summary, it may be useful to review the issues that augur confrontation. First is the matter of nation-states as controlling entities of social existence. To what extent will they or can they—given the different dominant forces at work within them—yield some of their sovereignty to larger, shared endeavors with which they can then interact in a movement toward a higher form of social cooperation and collectivity? Will the technology of our time, with its wide range of capabilities—from the threat of mutual nuclear destruction to the promise of space-transmitted global education and self-enlightment—augment or diminish the chances of amending the nation-state structures?

A second issue is, in a sense, a negative answer to the preceding question. Instead of stimulating an equitable internationalism of negotiated stability, the huge power and colossal cost of space-based telecommunication may further consolidate the world status of its present chief owners and operators. To compete in the launching of satellite systems is beyond the reach of most of the nations that are most in need of developmental and survival information systems. Why would those who have the know-how and the access want to share it equitably when such sharing may strengthen what they perceive as enemy or hostile forces? Accordingly, it may be anticipated that the imperatives of present market forces may stultify or control the direction of development of DBS in such a way as to prevent its uses for goals other than the expansion of markets and extension of Western ideology.

Within this conflict, which pits the trend toward international action against the driving force of superpower status, represented, in particular, by the world role of the United States, lies the awesome power of military technical prowess, subjecting civil society to ever-greater tension. That large parts of the new information system breakthroughs stem from military research and are under military control makes it ever more difficult to maintain and exercise the once-traditional civilian control processes of democracy.

Another issue that deepens the complexity of the United States role in world affairs is the relationship between private-enterprise corporations and the state. The development of DBS and its predecessors in telecommunications has illustrated the long history of the private corporate role as it has evolved from subservience, through integration and interdependence, to dominance. In the United States, both domestic policies and foreign relations have for more than a century been shaped and distorted to suit the needs of capitalist development rather than the needs of the whole society. That each discovery of new ways to better the condition of humankind must be exploited only to the degree that it can be made to realize a profit for invested capital has become part of the controlling ideology, now so deeply embedded that it goes almost unchallenged. The fact that this political and economic system has effected high living standards for large segments of the population has made it easier to conceal the ways in which it has sucked resources and gains from other segments of the population and from large portions of the rest of the world, particularly from those countries and peoples least able to resist. The possibility of altering or reversing this domination exists, but prolonged confrontation is to be expected, both in such international bodies as the United Nations and its rule-making specialized agencies like the ITU as well as in the insurgencies of guerrilla warfare and self-managed revolutions that are helped in important ways by radio communication, operating clandestinely or with the starved technology of a Third World economy.

Finally, there is the issue of ideology, which shows itself so openly in the information wars that use all known techniques of persuasion to win over minds and hearts. Here, anti-Communism as justification plays its continuing role to build up this country's open television and radio campaign against the socialist countries, a campaign conducted both by the government and by the huge media industry. A parallel, but less known, undertaking is the vast media penetration conducted by the CIA, with telecommunications operations that are directly linked to military action, as in Nicaragua and Afghanistan. On a world scale, Pentagon satellite link-ups make accurate surveillance and effective intervention ever more feasible.

This study has described the historical development of the geopolitical, socioeconomic, and, in certain respects, technological dominance of the United States in relation to international telecommunications. At the same time, it has outlined possible forms of resistance, some of which may be observed in different arenas of action. Such transformational processes evoke the promise, or at least the potential, of genuine improvement in the lives of those long denied, including the hundreds of millions in the Third World. But the use of the full potential of DBS to meet the developmental needs of humankind will not be easily accomplished before the advent of a far more comprehensive restructuring of the world order.

Glossary

Analog transmission The transmission of a continuously variable signal as opposed to a discretely variable signal. (Physical quantities such as temperature are continuously variable and so are described as "analog.")

ARABSAT The communications satellite consortium of twenty-two Arab states.

Bandwidth The size of the range of frequencies that are utilized in sending transmissions. The greater the bandwidth, the greater the number of channels it can carry and the greater the probability of transmitting data without error or distortion.

Beam A satellite antenna may focus the transmitted power signal to desired "beam" shapes: wide, to cover the visible hemisphere; irregular, to fit specific geographic areas; narrow, to correspond to particular receiving targets.

Broadcast satellite service (BSS) The ITU designation for satellite transmissions intended for designated earth stations for distribution to individual receivers.

Communication satellite An artificial space satellite designed to act as a telecommunications microwave relay. Most communications satellites are launched by rocket into an orbit 22,300 miles above the equator and are set at a velocity that is synchronized with that of the rotation of the earth so that they maintain what appears to be a stationary position relative to the earth (the geostationary orbit). They are solar powered, but have a seven-year "life expectancy," set by the amount of fuel they can carry to use for repositioning to maintain the correct orbit.

Communication satellite system Includes the operational (and spare) satellites, manned earth stations for tracking and controlling satellites, and a large number of earth stations serving the communication network.

COMSAT Communications Satellite Corporation, a quasi-public U.S. company established in 1962 by Congress in close alliance with major telecommunications corporations. In 1964, it created Intelsat, in which it is the U.S. signatory and for which it originally provided technical and operational management services. As COMSAT has grown, it has emerged as a major domestic competitor of other telecommunications corporations (operating through a separate subsidiary).

COPUOS United Nations Committee on the Peaceful Uses of Outer Space

Digital signal A discontinuous signal; its various states are discrete intervals apart. Digital techniques increase capacity relative to present analog techniques by a factor of close to 5.

Direct broadcast satellite (DBS) Direct broadcast by satellite is satellite broadcasting intended for direct reception by individual receiving sets.

Earth station Ground-based equipment, including the dish antenna used to receive or send-and-receive signals via satellite.

"Eclipse time" The dark period when the shadow of the earth comes between a satellite and the sun. Since the satellite is solar powered, the "eclipse time"

should occur after midnight to permit a recharging of the batteries before the start of the next broadcasting period.

EUTELSAT The communications satellite consortium of European nations.

Federal Communications Commission (FCC) The government regulatory agency established by Congress in 1934 and charged with regulating the telecommunications industry.

Fiber-optic transmissions Thin filaments of glass or other transparent materials through which a light beam can be transmitted for long distances by means of multiple internal reflections.

Fixed Satellite Service (FSS) The ITU term for defining point-to-point prearranged satellite transmissions; an example is a television network that transmits signals via satellite to a cable company for subsequent distribution by cable to subscribers.

Frequency The rate at which an electromagnetic (radio) wave alternates. It is measured in terms of how many waves or cycles occur per second. One cycle per second is a hertz (named after the physicist Heinrich Hertz, whose work in the 1880s showed that electromagnetic waves occur as a result of rapid electrical oscillations). One thousand waves per second is a kilohertz; 1 million, a megahertz; 1 billion, a gigahertz.

Frequency bands Defined arbitrarily as follows:

Range	*Term*
Below 3 megahertz	Low frequency (LF)
3–30 megahertz	High frequency (HF)
30–300 megahertz	Very-high frequency (VHF)
300 megahertz–3 gigahertz	Ultrahigh frequency (UHF)
3–30 gigahertz	Extremely high frequency (EHF)

Geostationary orbit (GSO) See *Communication satellite*.

INSAT The communications satellite system of India.

Intelsat International Telecommunications Satellite Consortium. It was formed by COMSAT in 1964 as a global communication system in which nation-states become members by buying stock. It currently has 110 members, owns 16 satellites, and provides service for most of international telecommunications.

International Telecommunication Union (ITU) A specialized agency of the United Nations, established originally in 1865 by international treaty as the International Telegraph Union. By periodic convention and conference, its approximately 160 member states agree on standardized communications procedures and practices, including frequency allocations and radio regulations.

Intersputnik International System and Organization of Space Communications. Formed by Soviet initiative in 1970, it has become a global system and currently has fourteen member states that lease capacity on Soviet satellites over the Atlantic and Indian oceans.

Ionosphere The ionized layers in the upper atmosphere, between 70 and 250 miles from the earth's surface, which reflect shortwave radio signals back to earth, thus permitting long-distance radio communication, but subject to static.

Microwave Electromagnetic waves in the microwave frequencies spectrum above 890 megahertz (UHF and EHF). Unlike longer wave radio signals (including "shortwave"), microwave is not reflected or affected by the ionosphere. It requires line-of-sight transmission. Signals from satellites are transmitted in microwave frequencies above 3,700 megahertz (3.7 gigahertz).

PALAP The communication satellite system of Indonesia.

Point-to-point A connection permanently established between two specific stations.

Radio communication Any telecommunication by means of radio waves.

Radio waves Electromagnetic waves of frequencies between 10 kilohertz and 3 megahertz, propagated without guide in free space.

Satellite signals Transmitted from earth station on an up-link frequency; amplified by transponders on the satellite; retransmitted to ground station on a different (down-link) frequency.

Scrambler A coding device applied to a digital channel that produces an apparently random bit sequence. A corresponding device is used to decode the channel; that is, the coding is reversible.

Spectrum A continuous range of frequencies usually wide in extent, within which waves have some specific common characteristics. In that portion of the electromagnetic spectrum utilized for telecommunications, specific uses are related to the characteristics of each segment as follows:

Band designation	*Uses*
Very-low frequency	Voice frequencies on telephone nonmultiplexed circuits
Low frequency	Suboceanic cables, long-wave radio
Medium frequency	AM-radio broadcasting, land-based coaxial cable
High frequency	Shortwave radio
Very-high frequency	Walkie-talkie, vehicular mobile radio
	FM-radio broadcasting
	TV broadcasting
Ultrahigh frequency	TV broadcasting, microwave radio, satellite-relayed radio

Telecommunication Any process that permits the passage of information from a sender to one or more receivers in any usable form (printed copy, fixed or moving pictures, visible or audible signals) by means of any electromagnetic system (electrical transmission by wire, radio, optical transmission, guided waves). It includes telegraphy, telephone, video-telephony, and data transmission.

Transponder A receiver–transmitter combination that retransmits the received signal, greatly amplified and at a different frequency. Communication satellites usually contain several transponders.

Appendix A: Highlights of International Broadcasting History

1831	Faraday discovers electrical induction
1835	Morse designs telegraph sender and key
1838	First commercial telegraph link in England
1844	First United States telegraph circuit
1861–65	United States Civil War
1865	First international treaty on telegraph regulation, Paris; twenty nations sign agreement and form International Telegraph Union
1866	Western Union, first privately owned national industrial monopoly, is established
1866	First transatlantic undersea cable
1875	International Telegraphy Union meeting, St. Petersburg
1876	Telephone is invented
1885	Berlin conference to apportion Africa among European colonializing powers
1895–99	Marconi (England) and Popov (Russia) conduct first wireless experiments
1903–06	First international conference on regulation of "telegraphy without wires," Berlin
1912	*Titanic* sinks
1912	Second international conference on wireless (radio), London
1914–18	World War I
1917	Russian Revolution
1919	Treaty of Versailles is signed; League of Nations is formed
1919	Soviet Russia begins international broadcasting
1920	Harding defeats Wilson; United States stays out of League
1920	Radio Corporation of America is formed
1920s	Successful demonstrations of television transmissions
1927	Congress enacts U.S. Radio Act of 1927
1927	Third international conference on radio, Washington, D.C.
1927	First commercial transatlantic radiotelephone
1929	Great Depression begins
1931–32	France and Great Britain begin shortwave international broadcasting service
1932	Fourth international conference on radio, Madrid; signatories of treaty on radio decide to merge with International Telegraph Union, and new agency is called the International Telecommunication Union (ITU)
1932	Roosevelt is elected president
1933	Hitler becomes chancellor of Germany; Nazis begin international broadcasting service
1939–45	World War II

1945 United States begins peacetime shortwave international broadcasting service, Voice of America
1945 United Nations is formed, with fifty-one member states
1947 International Telecommunication Union becomes a United Nations specialized agency at first postwar ITU conference, Atlantic City
1950 United States begins broadcasts of Radio Liberty and Radio Free Europe to Soviet Union and Eastern Europe
1954 First meeting of nonaligned nations, Bandung
1957 Soviet Union launches first space satellite, *Sputnik,* sending first radio signals from space
1958 United Nations establishes Committee on the Peaceful Uses of Outer Space (COPUOS)
1959 ITU conference makes first radio-frequency allocations for space communications
1960 United Nations membership reaches 100
1960 NASA launches *ECHO-1,* first telecommunications satellite
1962 Congress creates COMSAT, quasi-private corporation to provide satellite telecommunications services
1962 AT&T launches *TELSTAR I,* first privately owned active communications satellite
1963 ITU allocates segment of microwave band for use of communications satellites
1964 COMSAT sets up international consortium, Intelsat, to provide global telecommunications satellite services for its member states
1965 ITU conference, Montreux, considers satellite broadcasting; votes to exclude Union of South Africa
1965 First UNESCO conference on space communications
1965 Launch of first Intelsat *EARLY BIRD* communications satellite; Soviet Union launches first communications satellite, *Molniya*
1967 United States and the Soviet Union sign Treaty on Peaceful Uses of Outer Space
1968 First United Nations conference: "Exploration of Peaceful Uses of Outer Space"
1970 United Nations membership reaches 127
1971 ITU conference on space communications adopts Regulation 428A to prevent spillover of satellite broadcast signals into countries without their prior consent
1971 Soviet Union organizes Intersputnik, a satellite telecommunications network linking socialist countries
1972 General Assembly votes to assign COPUOS the task of developing guidelines, including prior consent, for international direct broadcasting by satellite (DBS)
1972 UNESCO adopts declaration of principles for satellite broadcasting, including the requirement that for direct satellite broadcasting there be prior agreement between the sending country and the receiving countries
1973 ITU conference, Malaga, Torremolinas, first asserts the rights of member states to "equitable access" to the space broadcasting spectrum; first assertion of registration jurisdiction over GSO
1974 United States launches first domestic commercial satellite, Western Union's *Westar I*

1976 UNESCO conference, Nairobi, endorses call for "new world information order"

1982 General Assembly votes to adopt guidelines for international direct television broadcasting by satellite (DBS), including the prior-consent provision

1982 Second United Nations conference on space: "Unispace '82" in Vienna

1983 United Nations observes International Communications Year

1984 United States withdraws from UNESCO

1985 United Nations celebrates fortieth anniversary; membership reaches 160

1985 ITU holds first of two conferences on more equitable sharing of the geostationary orbit

Appendix B: Highlights in the History of the International Telecommunication Union (ITU) and Its Forerunners

The International Telecommunication Union (ITU) is a specialized agency of the United Nations and is the oldest international regulatory agency. Its purpose is to develop and maintain cooperation in the international use of telecommunication services, including radio, telephone, and telegraph. The ITU is established and guided by the International Telecommunication Convention, which is subject to revision at periodic "plenipotentiary meetings" of the 161 member states. In addition, there are special ITU committees and conferences that develop policy affecting many aspects of international telecommunications. International broadcasting services are the concern of two ITU entities: the International Frequency Registration Board (IFRB) and the International Radio Consultative Committee (CCIR). Telephone and telegraph services are considered by the International Telephone and Telegraph Consultative Committee (CCITT).

1865 *Paris*. Austro-German Telegraph Union merges with West European Telegraph Union to form *International Telegraph Union*. Twenty member states formulate *International Telegraph Convention*, first international agreement to regulate rates and other aspects of wire telegraph services across national borders. No permanent organization is set up, reportedly because of unwillingness to give credit to Napoleon III as initiator.

1868 *Vienna*. Member states of International Telegraph Union set up permanent organization, establish secretariat (a staff of three, in the "Berne Bureau"), operated by the Swiss government.

1871–72 *Rome*. Nonvoting memberships are authorized and extended to ten private telegraph companies.

1875 *St. Petersburg*. United States sends observers but does not join. "Colonial voting" is authorized, allowing additional votes to member states with colonies.

1879 *London*. Twenty-four member states.

1885 *Berlin*. The organization undertakes development of regulations for international telephone communications.

1903 *Berlin*. Germany calls a Preliminary Conference Concerning Wireless Telegraphy. Protocol is drafted to apply to international uses of "telegraph without wires" (later to become known as radio). United States is one of nine countries participating.

1906 *Berlin*. Continuation of 1903 meeting. Twenty-nine countries attend. Resulting document is called *International Radiotelegraph Convention*. No permanent organization is formed. United States subsequently ratifies Convention.

1912 *London*. Meeting of signatories of 1906 Convention. Though there is no permanent organization, the working relationship of the International Radiotelegraph Convention signatories becomes known as the International Radiotelegraph Union (or International Radio Union). The Berne bureau of the International Telegraph Union is designated as the secretariat for the new organization.

1927 *Washington*. Second major conference of International Radio Union (that is, of the signatories of the *International Radiotelegraph Convention*). Soviet Union is denied permission to attend. Formation of an International Radio Consultative Committee (analogous to the International Telegraph Consultative Committee and International Telephone Consultative Committee of the International Telegraph Union); members of Committees include private operating companies and representatives of governments.

1932 *Madrid*. Merger of International Radiotelegraph Union with International Telegraph Union to form *International Telecommunication Union*.

1938 *Cairo*. First meeting of merged organization.

1947 *Atlantic City*. Postwar reactivation and major reorganization, accomplished in three conferences. Seventy-four countries attend. ITU becomes specialized agency of the United Nations. Secretariat is transferred from Berne to Geneva, and staff is "internationalized." Membership in ITU is limited to sovereign states only. "Colonial voting" disallowed for new members. Continuing status is provided for Radio Consultative Committee (CCIR) and soon-to-be-merged Telegraph and Telephone Consultative committees (CCITT). Creation of International Frequency Registration Board (IFRB).

1952 *Buenos Aires*. Plenipotentiary meeting. Reorganization completed. Eighty-three members attend. Makeup of IFRB made more representative of membership.

1959 *Geneva*. World Administrative Radio Conference (WARC 59). First allocation of frequencies for space activities (1 percent of the frequency spectrum is designated for such purposes).

1959 *Geneva*. Plenipotentiary meeting, to update the Buenos Aires convention. ITU purposes are broadened by assigning it responsibility for fostering telecommunications in developing countries. Membership reaches 101.

1963 *Geneva*. Special meeting, the Extraordinary Administrative Radio Conference to Allocate Frequency Bands for Space Radiocommunication Purposes (EARC). First ITU conference on space telecommunications. Allocations made for many kinds of space communication (total of about 15 percent of the spectrum is designated for such purposes).

1965 *Montreux*. Plenipotentiary meeting. Centennial of ITU. Republic of South Africa expelled from ITU. Membership reaches 129. Administrative Council expanded to include representatives of African and Asian countries.

1971 *Geneva*. World Administrative Radio Conference for Space Telecommunications (WARC-ST; the second ITU meeting on space communication.) Allocations designated for Broadcasting Satellite Service (BSS) and Fixed Satellite Service (FSS). Adoption of regulation 428A which seeks to ensure countries against unwanted "spillover" from satellite radiation. Adoption of provisions favoring equal "orbit/spectrum rights" and disallowing "permanent vesting" for members with a head start in space.

1973 *Malaga-Torremolinas*. Plenipotentiary meeting; 146 members. Adoption of closer ties with the United Nations. Special fund established to promote technical cooperation between member states. "Equitable access" to GSO and to spectrum is adopted as main objective. "Colonial voting" is abolished.

1977 *Geneva*. World Administrative Radio Conference for the Planning of the Broadcasting-Satellite Service (WARC-BS) in frequency bands 11.7–12.2 GHz (in regions 2 and 3) and 11.7–12.5 GHz (in region 1). Regional plans formulated for use of the Broadcasting Satellite Service in regions 1 (Europe and Africa) and 3 (Asia and Oceania). Effect of plans is to prevent the use of broadcasting satellites for transborder broadcasting without prior agreement and to guarantee orbital space to all affected nations. Plans for region 2 (the Americas) postponed to subsequent planning conference.

1979 *Geneva*. World Administrative Radio Conference (WARC-79). General conference on radio (first since 1959). Revision of standards, upward expansion of Frequency Allocations Table (from 275 GHz to 400 GHz).

1982 *Nairobi*. Plenipotentiary meeting. Updating of Malaga Convention, including strengthening of "equitable access" provisions; 160 members.

1983 *Geneva*. Regional Administrative Radio Conference (RARC-84) to adopt regional plan for use of the Broadcasting Satellite Service in region 2 (the Americas).

1985 *Geneva*. World Administrative Radio Conference on the Use of the Geostationary Orbit (WARC-ORB-1); first of two. Compromise agreement allots each country at least one orbital slot.

1988 WARC-ORB-2.

1989 Scheduled plenipotentiary.

Appendix C: Principles Governing the Use by States of Artificial Earth Satellites for International Direct Television Broadcasting

Adopted by the United Nations General Assembly, 10 December 1982 (U.N. Resolution No. 37/92)

A. Purposes and objectives

1. Activities in the field of international direct television broadcasting by satellite should be carried out in a manner compatible with the sovereign rights of States, including the principle of non-intervention, as well as with the right of everyone to seek, receive and impart information and ideas as enshrined in the relevant United Nations instruments.

2. Such activities should promote the free dissemination and mutual exchange of information and knowledge in cultural and scientific fields, assist in educational, social and economic development, particularly in the developing countries, enhance the qualities of life of all peoples and provide recreation with due respect to the political and cultural integrity of States.

3. These activities should accordingly be carried out in a manner compatible with the development of mutual understanding and the strengthening of friendly relations and co-operation among all States and peoples in the interest of maintaining international peace and security.

B. Applicability of international law

4. Activities in the field of international direct television broadcasting by satellite should be conducted in accordance with international law, including the Charter of the United Nations, the Treaty on Principles Governing the Activities of States in the Exploration and Use of Space, including the Moon and Other Celestial Bodies, of 27 January 1967, the relevant provisions of the International Telecommunication Convention and its Radio Regulations and of international instruments relating to friendly relations and co-operation among States and to human rights.

C. Rights and benefits

5. Every State has an equal right to conduct activities in the field of international direct television broadcasting by satellite and to authorize such activities by persons and entities under its jurisdiction. All States and peoples are entitled to and should enjoy the benefits from such activities. Access to the technology in this field should

be available to all States without discrimination on terms mutually agreed by all concerned.

D. International co-operation

6. Activities in the field of international direct television broadcasting by satellite should be based upon and encourage international co-operation. Such co-operation should be the subject of appropriate arrangements. Special consideration should be given to the needs of the developing countries in the use of international direct television broadcasting by satellite for the purpose of accelerating their national development.

E. Peaceful settlement of disputes

7. Any international dispute that may arise from activities covered by these principles should be settled through established procedures for the peaceful settlement of disputes agreed upon by the parties to the dispute in accordance with the provisions of the Charter of the United Nations.

F. State responsibility

8. States should bear international responsibility for activities in the field of international direct television broadcasting by satellites carried out by them or under their jurisdiction and for the conformity of any such activities with the principles set forth in this document.

9. When international direct television broadcasting by satellite is carried out by an international intergovernmental organization, the responsibility referred to in paragraph 8 above should be borne both by that organization and by the States participating in it.

G. Duty and right to consult

10. Any broadcasting or receiving State within an international direct television broadcasting satellite service established between them requested to do so by any other broadcasting or receiving State within the same service should promptly enter into consultations with the requesting State regarding its activities in the field of international direct television broadcasting by satellite, without prejudice to other consultations which these States may undertake with any other State on that subject.

H. Copyright and neighbouring rights

11. Without prejudice to the relevant provision of international law, States should co-operate on a bilateral and multilateral basis for protection of copyright and neighbouring rights by means of appropriate agreements between the interested States or the competent legal entities acting under their jurisdiction. In such co-operation they should give special consideration to the interests of developing countries in the use of direct television broadcasting for the purpose of accelerating their national development.

I. Notification to the United Nations

12. In order to promote international co-operation in the peaceful exploration and use of outer space, States conducting or authorizing activities in the field of international direct television broadcasting by satellite should inform the Secretary-General of the United Nations, to the greatest extent possible, of the nature of such activities. On receiving this information, the Secretary-General should disseminate it immediately and effectively to the relevant specialized agencies, as well as to the public and the international scientific community.

J. Consultations and agreements between States

13. A State which intends to establish or authorize the establishment of an international direct television broadcasting satellite service shall without delay notify the proposed receiving State or States of such intention and shall promptly enter into consultation with any of those States which so requests.

14. An international direct television broadcasting satellite service shall only be established after the conditions set forth in paragraph 13 above have been met and on the basis of agreements and/or arrangements in conformity with the relevant instruments of the International Telecommunication Union and in accordance with these principles.

15. With respect to the unavoidable overspill of the radiation of the satellite signal, the relevant instrument of the International Telecommunication Union shall be exclusively applicable.

Notes

Chapter 2

1. It was transmitted in Morse code, which was eventually adopted worldwide, an example of how the technology of the industrializing nations spread its influence. Morse had worked out the code by using dots and dashes for the letters of the Roman alphabet and by assigning the simplest dot–dash combinations to the most frequently used letters in the English language.
2. Patents stem from the king's power to grant privilege. They were systematized and reformed by a statute enacted during the reign of James I that did away with all such grants except for "new manufacture," in which case the "true and first inventor" could be granted exclusive rights. Significantly, a scientific idea or principle is not patentable unless it is "shown how to apply it so as to gain therefrom some material advantage." *Encyclopaedia Britannica,* 11th ed. (New York: Cambridge University Press, 1911), S.V. "Patent."
3. Article I, Section 8, Paragraph 8.
4. *The World Almanac and Book of Facts* (New York: Doubleday, 1978), p. 817.
5. Ibid.
6. Boyd C. Shafer, *Faces of Nationalism* (New York: Harcourt Brace Jovanovich, 1972), p. 29. Shafer points to three necessary conditions for the emergence of nations: (1) growth of the secular state, (2) weakening of the church and feudalism, and (3) development of communication.
7. Article I, Section 8, Paragraph 7.
8. An interesting history of the privatization of the telegraph is found in Richard B. Duboff, "The Rise of Communications Regulation: The Telegraph Industry, 1844–1880," *Journal of Communications* 34 (Summer 1984): 52–66.
9. Oliver W. Larkin, *Samuel F. B. Morse and American Democratic Art* (Boston: Little, Brown, 1954), p. 155.
10. Keith Clark, *International Communications, the American Attitude* (New York: Columbia University Press, 1931).
11. Ibid., p. 112.
12. Larkin, *Samuel F. B. Morse,* p. 141.
13. Gleason L. Archer, *History of Radio to 1926* (New York: American Historical Company, 1938), p. 38.
14. The first International Telegraph Convention was formulated at a meeting of delegates from twenty European nations held in Paris in 1865, which is the year given as the founding date of the International Telegraph Union. George A. Codding, Jr., and Anthony M. Rutkowski, *The International Telecommunications Union in a Changing World* (Dedham, Mass.: Artech House, 1982), and George A. Codding, Jr., "The Politicization of the International Telecommunications Union: Nairobi and After," in *Policy Research in Telecommunications,* ed. Vincent Mosco (Norwood, N. J.: Ablex, 1984).

15. Leslie Bennett Tribolet, *The International Aspects of Electrical Communications in the Pacific Area* (Baltimore: Johns Hopkins University Press, 1929), p. 12.
16. Quoted in ibid., pp. 10–11.
17. The success of the laissez-faire approach was hard won and seldom secure. Its protection and elaboration during the past century have been matters of concern for many of its proponents, including the radio historian Gleason L. Archer, who adds this remark to his account of Morse's early hope that he might sell his invention to the government: "Had the development of the telegraph industry been left to the Federal government it is probable that it would have been laid aside as a mere theory of science that under government largess had been tried and proved correct." Archer, *History of Radio,* p. 32.
18. U.S. Naval War College, *International Law Situations, with Solutions and Notes.* (Washington, D.C.: Government Printing Office, 1907), p. 140.
19. Quoted in Clark, *International Communications,* pp. 142–143.
20. Naval War College, *International Law Situations,* p. 140.
21. Clark, *International Communications,* p. 121.
22. Ibid.
23. Clark lists the other Western Hemisphere countries that refused to agree to the convention: Canada, Mexico, Cuba, Central American states, Peru, and Paraguay. The role of the United States in the politics of most of these countries during the past century no doubt was a factor in shaping their policies. Ibid, p. 103.
24. Ibid., p. 104.
25. Llewellyn White and Robert D. Leigh, *Peoples Speaking to Peoples: A Report on International Mass Communication for the Commission on Freedom of the Press.* (Chicago: University of Chicago Press, 1946), p. 17.
26. Ibid.
27. Archer, *History of Radio,* pp. 74–75.
28. Ibid., p. 76.
29. Quoted from the Final Protocol (1903) in Naval War College, *International Law Situations,* p. 146.
30. Quoted in ibid., p. 149.
31. Clark described the issue twenty-five years later as one of "freedom of the air," a phrase that does not appear to have been in existence in 1906, although it was to become a rallying cry in the years after World War II. Clark, *International Communications,* p. 219.
32. Ibid., p. 223.
33. Archer, *History of Radio,* p. 63.
34. James M. Herring and Gerald C. Gross, *Telecommunications: Economics and Regulation* (New York: McGraw-Hill, 1936), p. 78.
35. Frank J. Kahn, ed., *Documents of American Broadcasting* (New York: Appleton-Century-Crofts, 1972), p. 6.
36. Herring and Gross, *Telecommunications,* p. 88.
37. Clark, *International Communications,* p. 225.
38. The rule allowing extra votes to countries with colonial possessions was not abolished until the 1973 plenipotentiary session of the ITU. Codding, "The Politicization of the International Telecommunications Union."
39. Clark, *International Communications,* p. 175.
40. Kahn, ed., *Documents,* pp. 8–9.

41. Ibid, p. 20.
42. Sidney Rogerson, *Propaganda in the Next War*. (London: Geoffrey Books, 1938), p. 100.
43. A cable sent to President Wilson from Postmaster General A. S. Burleson on March 14, 1919, referred to in Ray Stannard Baker, *Woodrow Wilson and World Settlement,* 3 vols. (Garden City, N.Y.: Doubleday, 1922), Vol. 2, p. 470; vol. 3, p. 426.
44. Ibid., vol. 3, p. 426.
45. Ibid., vol. 2, p. 479.
46. Quoted in ibid., vol. 2, p. 483.
47. Quoted from a memorandum to President Wilson from Walter S. Rogers, in ibid., vol. 3, p. 432.
48. Ibid.
49. Ibid.
50. Archer, *History of Radio,* p. 167.
51. See Hugh G. J. Aitken, *The Continuous Wave: Technology and American Radio, 1900–1932* (Princeton, N.J.: Princeton University Press, 1985), and *Syntony and Spark: The Origins of Radio,* (New York: Wiley, 1975; Princeton, N.J.: Princeton University Press, 1985).
52. Archer, *History of Radio,* p. 157.
53. Ibid., p. 170.
54. Ibid.
55. Quoted in ibid., pp. 200–201.
56. In an admirable account of the prolonged and complicated negotiations leading to the creation of RCA, the historian Hugh G. J. Aitken reports almost reluctantly on some handwritten notes he found in Owen Young's private papers about the planned issuance of new common stock. The references in the notes are somewhat cryptic, but Aitken found one word to be unmistakable: *loot.* Aitken, *Continuous Wave,* p. 380.
57. Quoted from U.S. Congress, House Committee on Interstate and Foreign Commerce, *Hearings,* 67th Cong., 1st sess., May 10–13, 1921, pp. 112–116, in Tribolet, *International Aspects of Electrical Communications,* p. 44.
58. Archer, *History of Radio,* p. 232.
59. Quoted from the final agreement in ibid.
60. Tribolet, *International Aspects of Electrical Communications,* p. 61.
61. Ibid.
62. John D. Tomlinson, *The International Control of Radio Communications* (Ann Arbor, Mich.: Edwards, 1945), p. 57.
63. Ibid.
64. Tribolet, *International Aspects of Electrical Communications,* p. 11.
65. Tomlinson, *International Control of Radio Communications,* p. 48.
66. Ibid.
67. Baker, *Woodrow Wilson and World Settlement,* vol. 2, p. 487.
68. "President Coolidge's Message to Congress," December 7, 1926, in *Documents,* ed. Kahn, pp. 32–33.
69. Tomlinson, *International Control of Radio Communications,* p. 162. Only in 1956 was undersea cable developed to the point where it could transmit voice sound.
70. An international "juridical" conference of private radio entrepreneurs held in Paris in 1925 considered some of these issues and declared, among other things,

that "the air is free," perhaps an indication of a growing concern about further regulation. *New York Times,* 4 April 1925.

71. Clark, *International Communications,* pp. 226–227.
72. Ibid., p. 195.
73. Quoted in ibid.
74. Quoted in "Radio Congress to Determine Best Use for Wavelengths," *New York Times,* 9 October 1927, sect. 9, p. 17.
75. State Department comments on the agreement are in Clark, *International Communications,* p. 196. Additional details are in Tomlinson, *International Control of Radio Communications,* pp. 65–66.
76. Tomlinson, *International Control of Radio Communications,* p. 67.
77. Clark, *International Communications,* p. 230.
78. Joseph O. Chamberlain, "Radio Act of 1927," *American Bar Association Journal,* June 1927, quoted in ibid. Chamberlain wrote that "on the basis of the Radio Act this country can enter into the necessary international engagements to protect the American share in what is necessarily a world method of the distribution of ideas. . . . If the air is to be used successfully by radio, it must be on the basis of a world utility, regulated by a world public service commission through agreement of the governments."
79. Quoted in Tomlinson, *International Control of Radio Communications,* p. 159.
80. Ibid., pp. 107–108.
81. Ibid., p. 108.
82. Article 51 of the convention states: "Members and Associate members retain their entire freedom with regard to military radio installations of their army, naval and air forces." Harold K. Jacobson, "The International Telecommunication Union: ITU's Structure and Function," in *Global Communications in the Space Age: Toward a New ITU* (New York: John and Mary R. Markle Foundation and the Twentieth Century Fund, 1972), pp. 51–52.
83. A brief chronology of highlights in ITU history may be found in Appendix B.
84. Tomlinson explains that shortwave broadcasting enlarged the spectrum, but for technical reasons, it could not substitute for the medium and long waves. Tomlinson, *International Control of Radio Communications,* p. 152.
85. An account of how these problems were dealt with is in ibid., pp. 154ff.
86. Another interesting sidelight is the history of the radio series "Radio-Nations," planned in 1929 in cooperation with the Swiss government and inaugurated by the League of Nations in 1932. Its broadcasts included Assembly meetings, reports of commission activities, and such. Planned for both shortwave and long-wave transmissions, it was able to use only shortwave, which sharply limited its audience; it "had to stop its regular programmes at the beginning of hostilities—at the very moment when its voice was more than ever necessary." Arno Huth, *Radio Today: The Present State of Broadcasting,* (Geneva: Geneva Research Center, League of Nations, 1942), p. 137.
87. Quoted from the *League of Nations Assembly Journal* (1931), p. 115, in Tomlinson, *International Control of Radio Communications,* p. 227.
88. Quoted from the minutes of the conference in ibid., p. 229.
89. Ibid.
90. Ibid., p. 230.
91. Ibid., p. 233.
92. *New Times,* August 1972, p. 29, and N. S. Biryukov, *Television in the West and Its Doctrines* (Moscow: Progress Publishers, 1981), p. 196.

Chapter 3

1. George Norman Clark, *Unifying the World* (London: Swarthmore Press; New York: Harcourt Brace and Howe, 1920), pp. 10–11.
2. Ibid., p. 11.
3. Ibid., p. 70.
4. Ibid., p. 72.
5. Ibid.
6. Walter Lippmann, "The World Outside and the Pictures in Our Heads," in *Images of Man: The Classic Tradition in Sociological Thinking,* ed. C. Wright Mills (New York: Braziller, 1960), pp. 21–47.
7. Ibid., p. 33.
8. Ibid., p. 34.
9. Ibid., p. 35.
10. Ibid., p. 36.
11. Ibid.
12. Max Weber, "Science as a Vocation," in *From Max Weber: Essays in Sociology,* ed. H. H. Gerth and C. Wright Mills (New York: Oxford University Press, 1946), pp. 129–159.
13. Lippmann, "The World Outside," p. 44.
14. Harold D. Lasswell, *Propaganda Technique in the World War* (New York: Knopf, 1927), p. 5.
15. Ibid., p. 191.
16. Ibid., p. 122.
17. Harold D. Lasswell, "The Theory of Political Propaganda," in *Reader in Public Opinion and Communication,* ed. Bernard Berelson and Morris Janowitz (Glencoe, Ill.: Free Press, 1950), p. 177.
18. Lasswell, *Propaganda Technique,* p. 222.
19. Lasswell expanded the definition in his book *Politics: Who Gets What, When, How* (New York: McGraw-Hill, 1936).
20. Harold D. Lasswell and Dorothy Blumenstock, *World Revolutionary Propaganda: A Chicago Study* (New York: Knopf, 1939).
21. Ibid., p. 11.
22. Ibid., p. 300.
23. Ibid., p. 318.
24. Ibid., p. 341.
25. Ibid., p. 46.
26. Harold D. Lasswell, Ralph D. Casey, and Bruce Lannes Smith, *Propaganda and Promotional Activities: An Annotated Bibliography* (Minneapolis: University of Minnesota Press, 1935; Chicago: University of Chicago Press, 1969).
27. Ibid., p. 27.
28. Ibid., p. 21.
29. Ibid., p. 23.
30. Malcolm Macdonald Willey and Stuart A. Rice, *Communications Agencies and Social Life,* Recent Social Trends Series (New York: McGraw-Hill, 1933).
31. For a general account of the Rockefeller Foundation's interest in the study of radio in a context of ensuring economic stability, see Raymond Blaine Fosdick, *The Story of the Rockefeller Foundation* (New York: Harper, 1952), pp. 203–204, and Edward H. Berman, *The Ideology of Philanthropy: The Influence of the Carnegie, Ford, and Rockefeller Foundations on American Foreign Policy* (Albany: State University of New York Press, 1983).

32. Willey and Rice, *Communications Agencies,* p. 186.
33. Ibid., p. 195.
34. Ibid., p. 204–205.
35. Ibid., p. 209.
36. Ibid.
37. Oscar W. Riegel, *Mobilizing for Chaos: The Story of the New Propaganda* (New Haven, Conn.: Yale University Press, 1934), p. 37.
38. Ibid., p. 17. Riegel tended to see nationalism as a menacing force characteristic of such countries as Germany, Italy, and the Soviet Union. When evident in the United States, as with the government's "sympathy and cooperation" in helping create RCA as "100 per cent national," it was approvably benign.
39. Ibid., p. 21.
40. Ibid., p. 61. Riegel may have meant the Washington agreement of 1927, which left unresolved the problem of interference that would likely occur when U.S.–based transatlantic shipping companies used the medium-range frequencies already in use by European countries.
41. Ibid.
42. Ibid., pp. 64–66. This appears to be one of the earliest uses of the phrase "free flow," which was to become such a byword in later contestations over DBS and a "new world information order."
43. Rudolf Arnheim, *Radio* (1936; reprint, New York: Arno Press, 1971), p. 285.
44. Ibid., p. 226.
45. Ibid., pp. 227–228.
46. Ibid., p. p. 229.
47. Ibid., p. 234.
48. Ibid., pp. 234–235.
49. Ibid., p. 238.
50. Ibid., p. 239.
51. Ibid., pp. 240–241.
52. The *New York Times Index* for 1932 does not mention this development. Hadley Cantril and Gordon W. Allport, *The Psychology of Radio* (New York: Harper and Brothers, 1935), note that "the Socialist station, WEVD," won license renewal in 1931 after "considerable comment by the liberal press" (p. 51). They cite Federal Radio Commission Examiner's Report, No. 176, 1931: Docket No. 919, *Journal of Radio Law* 2 (1932): 119.
53. Arnheim, *Radio,* p. 245.
54. Emile Durkheim, *The Division of Labor in Society* (1896; reprint, New York: Free Press, 1933), p. 131.
55. Arnheim, *Radio,* p. 246.
56. Melvin L. DeFleur, *Theories of Mass Communication* (New York: McKay, 1966), p. 124, citing Harld D. Lasswell, "The Structure and Function of Communication in Society," in *The Communication of Ideas: A Series of Addresses,* ed. Lyman Bryson (New York: Harper and Brothers, 1948), pp. 37–51.
57. Cantril and Allport, *Psychology of Radio,* p. 48.
58. Ibid., p. 19.
59. Ibid., p. 21.
60. Ibid., p. 22.
61. Ibid., p. 24.
62. Ibid., p. 59.

63. Ibid., p. 270.
64. Ibid., p. 269.
65. Martin Jay, *The Dialectical Imagination: A History of the Frankfurt School and the Institute for Social Research, 1923–1950* (Boston: Little, Brown, 1973), and Susan Buck-Morss, *The Origin of Negative Dialectics* (New York: Free Press, 1977), pp. 153, 166.
66. Paul Lazarsfeld, *Radio and the Printed Page: An Introduction to the Study of Radio and Its Role in the Communication of Ideas* (New York: Duell, Sloan, and Pearce, 1940; New York: Arno Press, 1972), p. viii.
67. Ibid., p. xvii.
68. Ibid., p. 332.
69. DeFleur, *Theories of Mass Communication,* p. 176.
70. Herbert Blumer, "Elementary Collective Behavior," in *New Outline of the Principles of Sociology,* ed. Alfred McClung Lee (New York: Barnes and Noble, 1939), p. 186.
71. Ibid., p. 187.
72. Ibid., p. 193.
73. For example, Robert K. Merton, *Mass Persuasion: The Social Psychology of a War Bond Drive* (New York: Harper, 1946); Zechariah Chafee, Jr., *Government and Mass Communications,* 2 vols. (Chicago: University of Chicago Press, 1947); Wilbur Schramm, ed., *Mass Communications* (Urbana: University of Illinois Press, 1948): Joseph T. Klapper, *The Effects of Mass Media* (New York: Columbia University Press, 1949); and Carl I. Hovland, Arthur A. Lumsdaine, and Fred D. Sheffield, eds., *Experiments in Mass Communication,* Stouffer Studies in Social Psychology in World War II (Princeton, N.J.: Princeton University Press, 1950).
74. Talcott Parsons, "Propaganda and Social Control," in *Essays in Sociological Theory* (Glencoe, Ill.,: Free Press, 1949), p. 304.
75. Ibid., p. 305.
76. Harold D. Lasswell, *Democracy Through Public Opinion* (Menasha, Wis.: Banta, 1941), p. 165.
77. Ibid., p. 123.
78. Ibid., p. 165.
79. Ibid., p. 127.
80. Lasswell died in 1978. His posthumous publications include Harold D. Lasswell, Daniel Lerner, and Hans Speier, *Propaganda and Communication in World History,* 3 vols. (Honolulu: University of Hawaii Press, 1979), and Myres Smith McDougal, Harold D. Lasswell, and Lung-chu Chen, *Human Rights and World Public Order: The Basic Policies of an International Law of Human Dignity* (New Haven, Conn.: Yale University Press, 1980).
81. Lasswell, *Democracy Through Public Opinion,* p. 167.
82. Paul Lazarsfeld, "An Episode in the History of Social Research: A Memoir," in *The Intellectual Migration: Europe and America, 1930–1960*, ed. Donald Fleming and Bernard Bailyn (Cambridge, Mass.: Harvard University Press, Belknap Press, 1969).
83. Paul Lazarsfeld and Frank Stanton, eds., *Communications Research: 1948–1949* (New York: Harper and Brothers, 1949), p. xv.
84. Alex Inkeles, "Domestic Broadcasting in the USSR," in ibid., pp. 223–289.
85. For example, Nathan Leites, *A Study of Bolshevism* (Glencoe, Ill.: Free Press, 1953); Hadley Cantril, *Soviet Leaders and Mastery over Man* (New Bruns-

wick, N.J.: Rutgers University Press, 1960); Mark W. Hopkins, *Mass Media in the Soviet Union* (Pegasus, N.Y.: Western, 1970); Gayle Durham Hollander, *Soviet Political Indoctrination: Developments in Mass Media and Propaganda Since Stalin* (New York: Praeger, 1973); Burton Paulu, *Radio and Television Broadcasting in Eastern Europe* (Minneapolis: University of Minnesota Press, 1974); and Ellen Propper Mickiewicz, *Media and the Russian Public* (New York: Praeger, 1981).

86. CBS funded Klapper's book, and he later became head of the CBS research office, succeeding Frank Stanton, who became president of CBS. *The Effects of Mass Media* (New York: Columbia University Press, 1949) was published in a slightly different form eleven years later as *The Effects of Mass Communication: An Analysis of Research on the Effectiveness and Limitations of Mass Media in Influencing the Opinions, Values, and Behavior of Their Audiences* (New York: Free Press, 1960).
87. Klapper called these "needs," a term that came into wide use in communications research; it is suggestive of supply-and-demand market relationships.
88. Klapper, *Effects of Mass Media,* pp. 252–253.
89. Ibid., p. 51.
90. Lyman Bryson, ed., *The Communication of Ideas, a Series of Addresses* (New York: Harper and Brothers, 1948), p. 7.
91. Lasswell, "The Structure and Function of Communication in Society," in ibid., p. 42.
92. Merton, *Mass Persuasion.*
93. Paul Lazarsfeld and Robert K. Merton, "Mass Communication, Popular Taste and Organized Social Action," in *Communication of Ideas,* ed. Bryson, p. 96.
94. Charles Siepmann, "Radio," in *Communication of Ideas,* ed. Bryson, p. 181. Siepmann was a chief author of the FCC report, which was issued in 1946 and is reprinted in Frank J. Kahn, ed., *Documents of American Broadcasting* (New York: Appleton-Century-Crofts, 1972), pp. 125–206.
95. Kahn, ed., *Documents,* p. 125.
96. Lazarsfeld and Merton, "Mass Communication," p. 104.
97. Ibid., p. 114.
98. Ibid., p. 116.
99. Ibid., p. 117.
100. Robert K. Merton, *Social Theory and Social Structure: Toward the Codification of Theory and Research* (Glencoe, Ill.: Free Press, 1949), p. 5.
101. Ibid.
102. Ibid., p. 15.
103. Ibid., p. 16.
104. Ibid., p. 68.
105. Ibid., p. 206.
106. Ibid., p. 211.
107. Ibid.
108. Ibid., p. 213.
109. Ibid., p. 216.
110. Berelson and Janowitz, eds., *Reader in Public Opinion and Communication,* p. ix.
111. Ibid.
112. Quoted in Ralph O. Nafziger, ed., *Introduction to Mass Communications Research* (Baton Rouge: Louisiana State University Press, 1958), p. 16.

113. Bruce L. Smith, Harold D. Lasswell, and Ralph D. Casey, *Propaganda, Communication, and Public Opinion: A Comprehensive Reference Guide* (Princeton, N.J.: Princeton University Press, 1946).
114. Bruce Lannes Smith and Chitra M. Smith, eds., *International Communication and Political Opinion: A Guide to the Literature* (Princeton, N.J.: Princeton University Press, 1956). The original bibliography was reissued thirteen years later: Harold D. Lasswell, Ralph D. Casey, and Bruce Lannes Smith, *Propaganda and Promotional Activities: An Annotated Bibliography* (Chicago: University of Chicago Press, 1969). The introduction, "What Next?" was written by Lasswell.
115. Smith and Smith, eds., *International Communication and Political Action,* p. 7.
116. Ibid., p. 11.
117. Ibid., p. 12.
118. Ibid., p. 14.
119. Leonard W. Doob, *Communication in Africa: A Search for Boundaries* (New Haven, Conn.: Yale University Press, 1961). The book was funded by the Quartermaster Research and Engineering Command, United States Army.
120. Wilson P. Dizard, *The Strategy of Truth: The Story of the U.S. Information Service* (Washington, D.C.: Public Affairs Press, 1961). The chapter titles are revealing: "The Battle for World Opinion," "Words at War," and "The Delicate Art of Exporting Culture."
121. For example, Lucian W. Pye, ed., *Communications and Political Development* (Princeton, N.J.: Princeton University Press, 1963); Wilbur Schramm, *Mass Media and National Development: The Role of Information in the Developing Countries* (Stanford, Calif.: Stanford University Press, 1964, 1967); and Lewis Anthony Dexter and David Manning White, eds., *People, Society, and Mass Communications* (New York: Free Press, 1964).
122. For example, P. H. Coombs, *The Fourth Dimension of Foreign Policy* (New York: Harper & Row, 1964); Harold D. Lasswell, *World Revolutionary Elites,* MIT Studies in Comparative Politics (Cambridge, Mass.: MIT Press, 1965); Lucian Pye and Sidney Verba, eds., *Political Culture and Political Development* (Princeton, N.J.: Princeton University Press, 1965); and Walter Phillips Davison, *International Political Communication* (New York: Praeger, 1965).
123. Davison, *International Political Communication,* p. 5.
124. Ibid., p. 8.
125. Published in *Counter-Spy* (May–July 1981): 11–12. (Now called *National Reporter,* Box 21279, Washington, D.C.)
126. Quoted in George Gerbner, *Mass Media Policies in Changing Cultures* (New York: Wiley, 1977), p. 47.
127. Pye, ed., *Communications and Political Development,* p. 8.
128. Ibid.
129. Ibid., pp. 12–13.
130. Ithiel de Sola Pool, "The Mass Media and Their Interpersonal Social Function in the Process of Modernization," in *People, Society, and Mass Communications,* ed. Dexter and White, p. 433.
131. Ibid., p. 434. Pool argued that "Indian logic" would never accept this law, since Indian logic sees the possibility of statements being both true and false simultaneously.
132. Ibid.

133. Ibid., p. 439.
134. Daniel Lerner, "Toward a Communication Theory" in *Communications and Political Development,* ed. Pye, p. 350.
135. Daniel Lerner, *The Passing of Traditional Society: Modernizing the Middle East* (Glencoe, Ill.: Free Press, 1958).
136. Cited in Ithiel de Sola Pool, "Direct Broadcast Satellites and Cultural Integrity," *Society* (September–October 1975): 48. By 1975, Pool had become a fellow of the American Academy of Arts and Sciences and a member of the Council on Foreign Relations.
137. Ibid.
138. Ibid., p. 54.
139. Ibid., p. 53.
140. Ibid., p. 55.
141. Ithiel de Sola Pool, "Content Analysis and the Intelligence Function," in *Politics, Personality and Social Science in the Twentieth Century: Essays in Honor of Harold D. Lasswell,* ed. Arnold A. Rogow (Chicago: University of Chicago Press, 1969), p. 197.
142. Daniel Lerner, "Managing Communication for Modernization: A Developmental Construct," in *Politics, Personality and Social Science in the Twentieth Century,* ed. Rogow, p. 171.
143. Ibid., p. 178.

Chapter 4

1. *Encyclopaedia Britannica,* 11th ed. (New York: Cambridge University Press, 1910), S.V. Vol. 14, p. 697.
2. Ibid, p. 698.
3. Ibid.
4. Ibid, p. 696.
5. Georg Schwarzenberger, *The Dynamics of International Law* (Milton, Eng.: Professional Books, 1976), p. 124.
6. Oliver James Lissitzyn, *International Law Today and Tomorrow* (Dobbs Ferry, N.Y.: Oceana Publications, 1965), p. 65.
7. Philip C. Jessup, *A Modern Law of Nations: An Introduction* (New York: Macmillan, 1948), p. 2.
8. G. I. Tunkin, *Theory of International Law* (Cambridge, Mass.: Harvard University Press, 1974), p. 49.
9. Ibid., pp. 49–87.
10. Ibid., p. 260.
11. Ibid., p. 243.
12. Ibid., p. 225.
13. Lissitzyn reports that between 1917 and 1957, the Soviet Union negotiated 2,516 treaties and agreements; of the treaties, most were with the United States. Lissitzyn, *International Law Today and Tomorrow,* p. 54. He adds in a footnote that "in 1963 the U.S. Department of State listed 128 treaties and agreements as being in force between the two states."
14. Korovin is described by C. Wilfred Jenks as an eminent international lawyer who in 1965 was the chairman of the Scientific Research Committee on Legal Problems of Outer Space of the Institute of Law of the USSR Academy of Sciences. C. Wilfred Jenks, *Space Law* (New York: Praeger, 1965), p. 133.

15. E. Korovin, "International Status of Cosmic Space," *International Affairs* [Moscow] (January 1959): 58.
16. Jenks, *Space Law,* p. 174.
17. Ibid.
18. Lincoln P. Bloomfield, "The Prospects for Law and Order," in *Outer Space: Prospects for Man and Society,* ed. Lincoln P. Bloomfield (Englewood Cliffs, N.J.: Prentice-Hall, 1962), p. 159.
19. In the 1980s, as satellite-communication technology broadened to include computer-generated signals, the issue of the extent of ITU jurisdiction over transmissions via satellite caused new controversy. Multinational corporations and banks, which make very extensive use of data flows via satellite between their various international branches, argued that such flows should be treated differently from international satellite telecommunication or should be removed from ITU jurisdiction altogether.
20. Lissitzyn, *International Law Today and Tomorrow,* p. 70.
21. Ibid., p. 68.
22. Edward McWhinney, *"Peaceful Coexistence" and Soviet-Western International Law* (Leyden: Sijthoff, 1964), Appendix p. 131.
23. Schwarzenberger, *Dynamics of International Law,* p. 127.
24. Philip C. Jessup and Howard J. Taubenfeld, *Controls for Outer Space and the Antarctic Analogy* (New York: Columbia University Press, 1959), p. 299.
25. Ibid., p. 274.
26. *Everyman's United Nations* (New York: United Nations, 1968), pp. 68–69.
27. UNESCO, *Communication in the Space Age: The Use of Satellites by the Mass Media* (Paris: UNESCO, 1968), pp. 155–156.
28. Lissitzyn feels that the tradition of private interest groups in the West is one of the chief sources of respect for international law: "The needs of private persons carrying on economic activity across national boundaries . . . serve to strengthen the role of international law in governmental policies." Lissitzyn, *International Law Today and Tomorrow,* pp. 64–65.
29. Salvador de Madariaga, *Spain: A Modern History* (New York: Praeger, 1958), p. 466.
30. McWhinney, *"Peaceful Coexistence,"* p. 20. According to Lincoln Bloomfield, the ICAO convention that was adopted in 1944 provided that neither piloted aircraft nor pilotless aircraft could be flown over states without their permission. Bloomfield, "Prospects for Law and Order," p. 156.
31. A parallel may be observed in the present controversy over satellite mass communications, with the United States again arguing for the development of satellite-transmitted radio and television signals unrestrained by the sovereignty of those states into which the signals would penetrate.
32. John Cobb Cooper, *Explorations in Aerospace Law: Selected Essays by John Cobb Cooper,* ed. Ivan A. Vlasic (Montreal: McGill University Press, 1968), p. 39.
33. Ibid., p. 7.
34. Ibid., p. 30.
35. Cooper approvingly cited a description of the evolving international law of the high seas, taken from an opinion written in 1826 by Justice Storey of the Supreme Court: "Upon the ocean, then, in time of peace, all possess an entire equality. It is the common highway of all, appropriated to the use of all; and no one can vindicate to himself a superior prerogative there. Every ship sails

there with the unquestionable right of pursuing her own lawful business without interruption; but whatever may be that business, she is bound to pursue it in such a manner as not to violate the rights of others." Ibid., p. 39.
36. Korovin, "International Status of Cosmic Space," p. 55.
37. Ibid., pp. 58–59.
38. A. S. Piradov, ed., *International Space Law* (Moscow: Progress Publishers, 1976), p. 22.
39. Ibid., p. 24.
40. Bloomfield, "Prospects for Law and Order," p. 159.
41. Ibid., p. 158.
42. Ibid., p. 159.
43. "To jam, one transmits a disruptive signal at the same frequency as the offending signal. The two signals meet at the point of reception, the jamming signal overpowering the objectionable signal or preventing intelligible reception." M. S. Geston, "Direct Broadcasting from Satellite: The Case for Regulation," *New York University Journal of International Law and Politics* (Spring 1970): 80.
44. Ibid., p. 81.
45. Charles Henry Alexandrowicz, *The Law of Global Communications* (New York: Columbia University Press, 1971), p. 139.
46. Ibid., p. 154.
47. Ibid., p. 155.
48. Ibid., p. 156.
49. Ibid.
50. Ibid., p. 157.
51. Korovin, "International Status of Cosmic Space," p. 56.
52. Gyula Gal, *Space Law* (Leyden: Sijthoff; Dobbs Ferry, N.Y.: Oceana Publications, 1969).
53. Ibid., p. 37.
54. Ibid., p. 142.
55. Ibid., pp. 142–143.
56. Ibid., p. 141.
57. Quoted in William R. Hawkins, "How to Give Away Your Future," *National Review,* 16 April 1982, p. 410.
58. U.N. General Assembly Resolution 2916 (XXVII), 1972.
59. J. E. S. Fawcett, *International Law and the Uses of Outer Space* (Manchester: Manchester University Press), p. 63.
60. S. Houston Lay and Howard J. Taubenfeld, *The Law Relating to Activities of Man in Space* (Chicago: University of Chicago Press, 1970), p. 135.
61. Alexandrowicz, *Law of Global Communications,* p. 36.
62. Ibid., p. 37.
63. Ibid., p. 35.
64. Ibid., p. 37.
65. Carl Q. Christol, *The International Law of Outer Space,* U.S. Naval War College International Law Studies (Washington, D.C.: Government Printing Office, 1966), pp. 407–411.
66. Significantly, Christol held that this justification rested not on the source of the radio signal or the medium through which it passes, but on the content and nature of the signal.
67. Yuri Kolossov, "Legal Consequences of 'Spillover' Resulting from Satellite

Direct Broadcasting," in *Space Law Perspectives: Commentaries Based on Volumes 1–15 (1952–1972) of the Colloquia on the Law of Outer Space,* ed. Mortimer D. Schwartz (South Hackensack, N.J.: Rothman, 1976), p. 73.

68. Ibid., p. 74.
69. Fawcett, *International Law and the Uses of Outer Space,* pp. 62–63.
70. M. S. McDougal, Harold D. Lasswell, and I. A. Vlasic, *Law and Public Order in Space* (New Haven, Conn.; Yale University Press, 1963), p. 276.
71. Ibid., p. 371.
72. Lay and Taubenfeld, *Law Relating to Activities of Man in Space,* p. 16.
73. Robert C. Aldridge, *First Strike! The Pentagon's Strategy for Nuclear War* (Boston: South End Press, 1983), chap. 9; Jack Manno, "The Military History of the Space Shuttle," *Science for the People* 15 (1983): 6–11, 30–32; *Arming the Heavens: The Hidden Military Agenda for Space, 1945–1995* (New York: Dodd, Mead, 1984); Michio Kaku and Daniel Axelrod, *To Win a Nuclear War: The Pentagon's Secret Weapon* (Boston: South End Press, 1987).
74. Jonathan F. Galloway, *The Politics and Technology of Satellite Communication* (Lexington, Mass.: Lexington Books, 1972), p. 263.
75. Ivan A. Vlasic, quoted in "Proceedings of Conference on The Legal Problems of International Communications," *University of Toronto Law Journal* 20 (1970): 322. He also recommended that DBS be introduced "as a world community project" (p. 323).
76. Quoted in Lay and Taubenfeld, *Law Relating to Activities of Man in Space,* p. 124.
77. Communications Satellite Act of 1962, Section 102a.
78. Quoted in Galloway, *Politics and Technology of Satellite Communication,* p. 266.
79. Bloomfield, "Prospects for Law and Order," p. 169.
80. Gal, *Space Law.* pp. 264–266.
81. V. S. Vereschetin, "Legal Questions of Satellite Telecommunications at the Washington Intelsat Conference," in *Space Law Perspectives,* ed. Schwartz, p. 17.
82. Ibid.
83. Marcellus S. Snow, "Intelsat: An International Example," *Journal of Communications* 30 (Spring 1980): 153. Snow regarded this outcome as a victory for the United States, and added: "Undoubtedly, American hegemony in synchronous rocketry exerted much of the influence leading to the ultimate outcome."
84. Richard R. Colino, "Reply to Mr. Vereschetin," in *Space Law Perspectives,* ed. Schwartz, pp. 246–247.
85. Mortimer Schwartz, "Commentary," in *Space Law Perspectives,* ed. Schwartz, p. 207.
86. Julian G. Verplaetse, "Observations on Telecommunications and Outer Space," in *Space Law Perspectives,* ed. Schwartz, p. 363.
87. M. S. Geston, "Direct Broadcasting from Satellite," p. 85.
88. Ibid., p. 93.
89. In the 1980s, ESA made possible the launching of communications satellites for Eutelsat, an organization created by and serving the post and telecommunications agencies of major European governments.
90. Kathryn M. Queeney, *Direct Broadcast Satellites and the United Nations* (Alphen aan den Rijn, Netherlands: Sijthoff and Noordhoff, 1978), p. 83.

91. Quoted in A. Berrada, *Frequencies for Broadcasting Satellites,* International Broadcast Institute Monograph (London: International Broadcast Institute, 1972), p. 3.
92. Ibid.
93. In U.N. Document A/AC. 105/109, 1973.
94. For a summary of mass media initiatives undertaken by the United Nations and UNESCO, see Hilding A. Eek, "Principles Governing the Use of the Mass Media as Defined by the United Nations and UNESCO," in *National Sovereignty and International Communications,* ed. Kaarle Nordenstreng and Herbert I. Schiller (Norwood, N.J.: Ablex, 1979), pp. 173–191.
95. Sean MacBride et al., *Many Voices, One World: Towards a New, More Just, and More Efficient World Information and Communication Order* (London: Kogan Page, 1980), pp. 288–290.
96. Another source is the English-language journals published in socialist countries: *Democratic Journalist* (Prague), *International Affairs* (Moscow), and *New Times* (Moscow).
97. Article IX(1), in U.N. Document A/8771, 1971.
98. Jennifer Freeman, "Toward the Free Flow of Information: Direct Television Broadcasting via Satellite," *Journal of International Law and Economics* 13 (1979): 343–345.
99. Article V, in U.N. Document A/8771, 1971.
100. U.N. Document A/AC/WG 3/L.4, 1973, in U.N. Document 105/107. Annex IV, 1974.
101. Details drawn from Nandasiri Jasentuliyana and Roy S. K. Lee, "Direct Satellite Broadcasting," in *Manual on Space Law,* ed. Nandasiri Jasentuliyana and Roy S. K. Lee (Dobbs Ferry, N.Y.: Oceana Publications, 1979), vol. 1, pp. 288–295. Jasentuliyana is on the professional staff of COPUOS. This book provides detailed accounts of the various legal aspects of the DBS controversy.

Chapter 5

1. F. S. Ruddy, "The Freedom of Information and Direct Broadcast Satellites" (Paper presented at the Sixteenth Colloquium on Outer Space Law, International Institute of Space Law, 1973). On that occasion, Ruddy also read from Aleksandr Solzhenitsyn's Nobel Prize acceptance speech: "Mankind has become one, . . . surpassing all barriers through international broadcasting and printing"; and "Suppression of information leads to entropy and total destruction. . . . The people know nothing intelligent about the rest of the earth and are prepared to go and trample it down in the holy conviction that they come as liberators."
2. The participants included lawyers Abram Chayes, Paul L. Laskin, John Lawrence Hargrove, and Monroe E. Price; social scientists Emile McAnany and Ithiel de Sola Pool; and Leonard H. Marks, former director of the United States Information Agency.
3. Aspen Institute for Humanistic Studies, *Control of the Direct Broadcast Satellite: Values in Conflict* (Washington, D.C.: Aspen Institute Publications, 1974), pp. 83–84.
4. Margaret A. Blanchard has recently published a study of their campaign: *Exporting the First Amendment: The Press–Government Crusade of 1945–1953* (White Plains, N.Y.: Longman, 1986).

5. Y. N. Zasursky, "'Free Flow of Information': The Cold War and Reducing International Tension," *Democratic Journalist* [Czechoslovakia], no. 3 (1974): 8. See also Herbert I. Schiller, *Communication and Cultural Domination* (White Plains, N.Y.: International Arts and Sciences Press, 1976), pp. 24–46, and "Freedom from the 'Free Flow,'" *Journal of Communication* 24 (Winter 1974): 110–117; and Schiller's "'Free Flow of Information' Aids U.S. Capitalism," *Our Right to Know* (New York: Fund for Open Information and Accountability, 1985), pp. 4–5.
6. Zasursky, "'Free Flow of Information,'" p. 8. As examples of the practice of "free flow," he pointed to Czechoslovakia in 1968 and to Chile from 1970 to 1973; in both countries, the "international information monopolies" had helped conduct a systematic information war against the legal government.
7. John L. Hargrove, "International Law and the Case for Cultural Protectionism," in Aspen Institute, *Control of the Direct Broadcast Satellite*, p. 89.
8. Ibid., p. 94.
9. Ibid., p. 89.
10. Ibid., p. 95.
11. Edward R. Finch, Jr., and Amanda L. Moore, "Outer Space Law and the Global Community," *The International Lawyer* 8 (October 1974): 762.
12. Quoted in "Introduction to Part 4: International Communication in Transition," in *National Sovereignty and International Communications*, ed. Kaarle Nordenstreng and Herbert I. Schiller (Norwood, N.J.: Ablex, 1979), p. 234.
13. Paul L. Laskin, "Background Paper," in *Communicating by Satellite* (New York: Twentieth Century Fund, 1969), p. 14. In 1970, the Twentieth Century Fund published a second report on the topic, written by Gorden Weil. It revealed a change of heart. Weil wrote: "A requirement of consent is unobjectionable if it means no more than the right of a country to participate without veto in negotiations—in the ITU or elsewhere—concerning systems that will illuminate its territory" (p. 19). The phrase "without veto" indicates the new perspective.
14. Abram Chayes and Paul Laskin, *Direct Broadcasting from Satellites: Policies and Problems*, Studies in Transnational Legal Policy, no. 7 (Washington, D.C.: American Society of International Law, 1975); Paul L. Laskin and Abram Chayes, "International Satellite Controversy," *Society* 12 (September–October 1975): 30–40; and Abram Chayes and Leonard Chazen, "Policy Problems in Direct Broadcasting from Satellites," *Stanford Journal of International Studies* 5 (June 1970): 4–21.
15. "United States Proposes Voluntary Principles on Direct Broadcasting by Satellite," *U.S. Department of State Bulletin*, 22 April 1974 p. 17.
16. Chayes and Laskin, *Direct Broadcasting from Satellites*, p. 25.
17. Ibid., p. 23.
18. Ibid., p. 26.
19. Ibid., p. 28. They added that in the *Radio Luxembourg* case in the 1930s, the United Kingdom had asserted "and partially vindicated" the right under international law to jam unwanted broadcasts.
20. Ibid., p. 33.
21. Thomas Buergenthal, "The Right to Receive Information Across National Boundaries," in Aspen Institute, *Control of the Direct Broadcast Satellite*, pp. 80–81.
22. Ibid., p. 83.

23. Ibid., p. 81.
24. Ibid., p. 80.
25. Seyom Brown, Nina W. Cornell, Larry L. Fabian, and Edith Brown Riess, *Regimes for the Ocean, Outer Space and Weather* (Washington, D.C.: Brookings Institution, 1978), p. 3.
26. Ibid., p. 2.
27. Ibid., p. 160.
28. Ibid., p. 161.
29. Ibid., p. 160.
30. Ibid., p. 161.
31. Ibid.
32. Ibid., p. 162.
33. Leo Gross, "Some International Law Aspects of the Freedom of Information and the Right to Communicate," in *National Sovereignty and International Communications*, ed. Nordenstreng and Schiller, pp. 195–211.
34. Ibid., p. 197.
35. Ibid., p. 199.
36. Ibid. In support of this principle, Gross cited the advisory opinion of the Permanent Court of International Justice in the Greco-Bulgarian *Communities* case.
37. Ibid., pp. 196, 203.
38. Quoted in ibid., p. 205.
39. Ibid.
40. Ibid., p. 206.
41. James J. Gehrig, "Broadcasting Satellites—Prospects and Problems," *Journal of Space Law* 3 (Spring and Fall, 1975): 38.
42. Leo T. Stull, quoted in U.S. Congress, Senate Committee on Aeronautical and Space Sciences, *Soviet Space Programs, 1971–1975: Goals and Purposes, Organization, Resource Allocations, Attitudes Toward International Cooperation and Space Law*, 94th Cong., 2d sess., 1976, vol. 2, p. 196.
43. Christian Patermann, "Applicable Law in Cases of Tort Damages Caused by Direct Broadcast Satellites," *Journal of Space Law* (Spring and Fall 1975): 3, 110.
44. Ruddy, "Freedom of Information and Direct Broadcast Satellites," p. 82.
45. Friedman Memorial Program, "The Control of Program Content in International Telecommunications." Two sessions: "A Discussion of General Principles" and "Direct Satellite Broadcasting and the Third World," *Columbia Journal of Transnational Law* 13 (1974): 40–81.
46. Friedman Memorial Program, "The Control of Program Content in International Telecommunications: A Discussion of General Principles" (moderator: Fred Friendly; discussants: Abram Chayes, Frank Shakespeare, Harry Olsson, and David Webster), p. 52.
47. Ibid., p. 53.
48. Ibid., p. 57.
49. Quoted in Brown, Cornell, Fabian, and Riess, *Regimes for the Ocean*, pp. 156–157.
50. "Will They Stop Our Satellites?," *New York Times*, 22 October 1972, quoted in ibid., p. 156.
51. Quoted in Jennifer Freeman, "Toward the Free Flow of Information: Direct Television Broadcasting via Satellite," *Journal of International Law and Economics* 13, no. 2 (1979): 352, n. 148.

52. Quoted in U.S. Congress, Senate Committee on Foreign Relations, Subcommittee on International Operations, *The Implications of International Communications and Information: Hearings*, 95th Cong., 1st sess., 1976, p. 191.
53. Quoted in ibid., p. 219.
54. Monroe E. Price, "First Amendment and Television Broadcasting by Satellite," *UCLA Law Review*, 23 (June 1976): 879–903.
55. Friedman Memorial Program, "The Control of Program Content in International Telecommunications: A Discussion of General Principles," p. 54.
56. Ibid., p. 57. Chayes cited the bull horn, the loudspeaker, and soliciting cases, including *Lovell* v. *City of Griffin*, 303 U.S. 444 (91938) and *Cantwell* v. *Connecticut* 310 U.S. 296 (1040), adding: "In these cases the community is, under our First at its extreme, authorized to regulate the power of someone to impose speech on unwilling listeners" (p. 57).
57. Price, "First Amendment and Television Broadcasting by Satellite," p. 891.
58. Ibid.
59. Ibid., p. 893.
60. Ibid., p. 896.
61. United Nations, Committee on the Peaceful Uses of Outer Space, *Report of the Committee on the Peaceful Uses of Outer Space*, G.A.O.R., 32nd sess., Supplement no. 20 (A/32/20) in *Manual on Space Law*, ed. Nandasiri Jasentuliyana and Roy S. K. Lee (Dobbs Ferry, N.Y.: Oceana Publications, 1979), vol. 1, pp. 293–294.
62. John H. Chapman and Gabriel I. Warren, "Direct Broadcast Satellites: The ITU, UN and the Real World," in *Annals of Air and Space Law*, ed. N. M. Matte (Toronto: Carswell, 1979), vol. 4, p. 427.
63. Quoted from 47 C.F.R. Paragraph 73.788 (a) 1973, in Price, "First Amendment and Television Broadcasting by Satellite," p. 897.
64. Ibid. ITU regulations allot segments for shortwave broadcasting to each country to assign internally as it wishes. In the United States, about 90 percent of the allotment is assigned to the Voice of America, which by law is excluded from the jurisdiction of the FCC.
65. Ibid., p. 898.
66. Quoted from 44 U.S.C. Paragraph 310 (a) (4) 1970, in ibid., p. 899.
67. Ibid., p. 900.
68. Ibid., pp. 900–901.
69. Ibid., pp. 898–899.
70. Ibid., p. 901.
71. Freeman, "Toward the Free Flow of Information," p. 354. Freeman also asked whether the First Amendment can be considered to have "an extraterritorial reach that extends to foreign broadcast or audiences," citing the Supreme Court decision in *United States* v. *Curtiss-Wright Export Corporation* 399 U.S. 304, 318 (1936), as relevant.
72. Quoted in Jan Busak, "Perspectives de la télévision et de la radiodiffusion directe par satellite," in *Space Law Perspectives: Commentaries Based on Volumes 1–15 (1957–1972) of the Colloquia on the Law of Outer Space*, ed. Mortimer D. Schwartz (South Hackensack, N.J.: Rothman, 1976), p. 141.
73. Chapman and Warren, "Direct Broadcast Satellites," p. 426.
74. Richard E. Butler, "World Administrative Radio Conference for Planning Broadcasting Satellite Service," *Journal of Space Law* 5 (Spring and Fall 1977): 95–96.

75. Ibid., p. 96.
76. Chapman and Warren, "Direct Broadcast Satellites," pp. 421–423. A good description of the complicated plan is found in this monograph.
77. Wilson P. Dizard, "The US Position: DBS and Free Flow," *Journal of Communications* 30 (Spring 1980): 165.
78. Quoted in Kathryn M. Queeney, *Direct Broadcast Satellites and the United Nations* (Alphen aan den Rijn, Netherlands: Sijthoff and Noordhoff, 1978), pp. 152–153.
79. Ibid., p. 152.
80. Ibid.
81. Nandasiri Jasentuliyana, "Regulations Governing Space Telecommunication," in *Manual of Space Law*, ed. Jasentuliyana and Lee, Vol. 1, p. 202.
82. Ibid., p. 205.
83. George Jacobs, technical advisor to the United States Board for International Broadcasting, quoted in U.S. Congress, House Committee on Government Operations, *International Broadcasting: Direct Broadcast Satellites: Hearings Before a Subcommittee,* 97th Cong., 1st sess., 1981, p. 22.
84. Carl Q. Christol, "The International Telecommunication Union and the International Law of Outer Space," *Proceedings of the Twenty-second Colloquium of International Institute of Space Law* (New York: American Institute of Aeronautics and Astronautics, 1978), p. 48.
85. Ibid. The phrase "equality weapon" is quoted from Jean d'Arcy, "Direct Broadcast Satellites and Freedom of Information," in *The International Law of Communications,* ed. Edward McWhinney (Dobbs Ferry, N.Y.: Oceana Publications, 1971), p. 168.
86. Dizard, "The US Position," pp. 157–168.
87. Ibid., pp. 163–166.
88. Andrzej Gorbiel, "The Legal Status of the Geostationary Orbit: Some Remarks," *Journal of Space Law,* 6 (Fall 1978): 171.
89. Ogunsola O. Ogunbanwo, *International Law and Outer Space Activities* (The Hague: Martinus Nijhoff, 1975), p. 183.
90. Ibid., p. 184. Ogunbanwo cited a number of sources that support the United States position.
91. Ibid., p. 185.
92. The Bogotá Declaration is reprinted in *Journal of Space Law* 6 (Fall 1978): 193–195.
93. Butler, "World Administrative Radio Conference," p. 97.
94. Gorbiel, "Legal Status of the Geostationary Orbit," p. 177.
95. D. Goedhuis, "Influence of the Conquest of Outer Space on National Sovereignty: Some Observations," *Journal of Space Law* 6 (Spring 1978): 40–41.
96. S. Neil Hosenball, "The United Nations Committee on the Peaceful Uses of Outer Space: Past Accomplishments and Future Challenges," *Journal of Space Law* 7 (Fall 1979): 112.
97. Ibid.
98. D. Goedhuis, "Events of Interest," *Journal of Space Law* 7, (Spring 1979): 64. The subject was assigned to the Space Law Committee of the IISL to work in conjunction with the Air Law Committee of the ICAO.
99. Queeney, *Direct Broadcast Satellites and the United Nations,* p. 24.
100. Ibid., p. 66.

101. Eileen Galloway, "Direct Broadcast Satellites and Space Law," *Journal of Space Law* 3 (Spring and Fall 1975): 110.
102. Letter included in House Committee on Government Operations, *International Broadcasting,* p. 85.
103. Quoted in ibid., pp. 46–47.
104. Chapman and Warren, "Direct Broadcast Satellites," p. 430.
105. House Committee on Government Operaations, *International Broadcasting,* pp. 41–51.
106. Quoted in ibid., pp. 50–51.
107. Ibid.
108. Quoted in ibid., p. 27. These kinds of transmitter stations are called "slave transmitters," according to a former head of the Voice of America, Frank Shakespeare, in Friedman Memorial Program, "The Control of Program Content in International Telecommunications: A Discussion of General Principles," p. 42.
109. Subrata K. Sarkar, "Requirements for Establishing a Broadcast Satellite Service" (Paper presented at the Sixteenth Colloquium on Outer Space Law, Institute of Space Law, 1973), p. 85.
110. Patermann, "Applicable Law in Cases of Tort Damages," pp. 111–112.
111. Shakespeare, in Friedman Memorial Program, "The Control of Program Content in International Telecommunications: A Discussion of General Principles," p. 42.
112. Ibid., p. 43.
113. *The World Almanac and Book of Facts* (New York: Newspaper Enterprise Association, 1983), pp. 497–582.
114. Shakespeare, in Friedman Memorial Program, "The Control of Program Content in International Telecommunications: A Discussion of General Principles," p. 56.
115. Ibid.
116. Friedman Memorial Program, "The Control of Program Content in International Telecommunications: Direct Satellite Broadcasting and the Third World" (discussants: Luiz Felipe de Seixas Corrêa, Nandasiri Jasentuliyana, Abram Chayes, Irwin Singer, and Josef C. Nicols), p. 73.
117. Ibid.
118. Queeney, *Direct Broadcast Satellites and the United Nations,* p. 21.
119. Ibid., pp. 22–23.
120. Finch and Moore, "Outer Space Law and the Global Community," p. 761.
121. Ibid.
122. Eilene Galloway, "Consensus Decisionmaking by the United Nations Committee on the Peaceful Uses of Outer Space," *Journal of Space Law* 7 (Spring 1979): 8. She was referring to the Outer Space Treaty (1967), the Agreement on Rescue of Astronauts (1968), and the conventions on registration of and damages caused by space objects (1973 and 1976).
123. Ibid., p. 11.
124. Christol, "The International Telecommunication Union," p. 51.
125. P. P. C. Haanappel, "The Stagnating Development of International Space Law and its Causes" (Paper presented at the Twenty-second Colloquium on the Law of Outer Space, International Institute of Space Law, 1979), p. 152.
126. Dizard, "The US Position," p. 162.

127. Ibid., p. 167.
128. House Committee on Government Operations, *International Broadcasting*, p. 47.

Chapter 6

1. Veikko Pietilä, "On the Scientific Status and Position of Communication Research" Institute of Journalism and Mass Communication (Tampere, Finland, 1977, Mimeographed), p. 72.
2. Herbert I. Schiller, *The Mind Managers* (Boston: Beacon Press, 1973), p. 5.
3. Quoted in Nicholas Garnham, "Contribution to a Political Economy of Mass Communication," *Media, Culture and Society* 1 (1979): 134.
4. Raymond Williams, *Marxism and Literature* (London: Oxford University Press, 1977), p. 136.
5. Walery Pisarek, "Marxist-Leninist Theory and Methodology of Media Research: Principles and Directives," *Marxist-Leninist Theory and Methodology of Media Research,* ed. Walery Pisarek and Thomasz Goban-Klas (Cracow: Press Research Center, 1978), p. 29.
6. Ibid., p. 24.
7. Kaarle Nordenstreng, "From Mass Media to Mass Consciousness," in *Mass Media Policies in Changing Cultures,* ed. George Gerbner (New York: Wiley, 1977), p. 276.
8. Armand Mattelart, "For a 'Class Analysis' of Communication," in *Communication and Class Struggle,* ed. Armand Mattelart and Seth Siegelaub (New York: International General, 1979), vol. 1, p. 52.
9. J. D. Bernal, *The Social Function of Science* (New York: Macmillan, 1939), p. 131.
10. Karl Marx, "Preface to *A Critique of Political Economy,*" *Dynamics of Social Change,* ed. Howard Selsam et al. (New York: International Publishers, 1970), p. 52.
11. Howard J. Sherman, "'Technology vis-à-vis Institutions': A Marxist Commentary," *Journal of Economic Issues* 8 (March 1979): 188.
12. Vincent Mosco and Andrew Herman, "Radical Social Theory and the Communications Revolution" (Paper presented at the Fourth Annual Conference on the Current State of Marxist Theory, University of Louisville, 1979), p. 3.
13. Pietilä, "On the Scientific Status and Position of Communication Research," pp. 45–46.
14. Mattelart, "For a 'Class Analysis' of Communication," p. 46.
15. Pietilä, "On the Scientific Status and Position of Communication Research," p. 50.
16. Mattelart, "For a 'Class Analysis' of Communication," p. 39.
17. Ibid.
18. Pisarek, "Marxist-Leninist Theory and Methodology of Media Research," p. 33.
19. Seth Siegelaub, "Preface: A Communication on Communication," in *Communication and Class Struggle,* ed. Mattelart and Siegelaub, p. 15.
20. Herbert I. Schiller, "Information for What Kind of Society?" (Papers for Schiller's course "The Political Economy of Mass Communications in the United States," University of California at San Diego, 1982), p. 227.
21. Consider this statement from Venezuelan President Carlos Andres Perez in 1976:

"Our people live submitted to an uncontrolled invasion of news which inculcates distant values that threaten our national identity. Such a fallacious concept of freedom of information, leaving absolute decision-making power about the information given to people in the hands of transnational news enterprises can only be termed unjust. The international press only collects information that undermines the image of our people . . . ignoring our struggles, our efforts and just claims for a system of international justice." Quoted in "The Old Information Order—A Concept Trapped in Time" (adapted from a report of the Instituto Latinoamericano de Estudios Transnacionales (ILET) in Mexico), NACLA *Report on the Americas* 16 (July–August 1982): 15.

22. Siegelaub, "Preface," p. 19.
23. Karl Marx, "Preface to *A Contribution to the Critique of Political Economy,*" in *The Marx-Engels Reader,* ed. Robert C. Tucker (New York: Norton, 1972), p. 4.
24. Peter Berger and Thomas Luckman, *The Social Construction of Reality* (Garden City, N.Y.: Doubleday, 1966), p. 6.
25. Sherman, "'Technology vis-à-vis Institutions,'" p. 181.
26. Herbert I. Schiller, *Communication and Cultural Domination* (White Plains, N.Y.: International Arts and Sciences Press, 1976), p. 6.
27. Schiller quotes a capitalist enthusiast remarking about the special quality of "information": "You can sell information over and over again and still have it." Herbert I. Schiller, *Who Knows: Information in the Age of the Fortune 500* (Norwood, N.J.: Ablex, 1981), p. 9.
28. An early illustration of how selling an idea can be related to selling a commodity is found in the following excerpt from a 1926 memo that Will Hays, then president of Motion Picture Producers and Distributors of America, Inc., sent to the Secretary of State: "Trade follows the film, rather than the flag . . . the screen acts as a voiceless salesman for the goods it pictures." Quoted in P. Bender, D. Halleck, and B. Summers, *"De Películas:* Proposal for a Film" (1985, Mimeographed).
29. Leonardo Acosta, "Mass Media and Imperialist Ideology," in *Communication and Class Struggle,* ed. Mattelart and Siegelaub, pp. 153–154. To illustrate how architecture can be both propaganda and technology, Acosta, a Cuban, cites such "development projects" as Brasilia and Chandigarth, both of which are "presented as models for the so-called Third World" (p. 154).
30. Cees Hamelink, "The New Information Order and the Industrial State," in *Peace and the Sciences* (Vienna: International Institute for Peace, 1978), p. 6. Hamelink gave as examples of the advertising operations "the sales of Barbie dolls to starving children in Recife," "'white is beautiful' skin cream to black Africans," and "CIA activities and the design of Chilean counter-revolution."
31. For example, Thomas L. McPhail, *Electronic Colonialism: The Future of International Broadcasting and Communication* (Beverly Hills, Calif.: Sage, 1981); Jeremy Tunstall, *The Media Are American: Anglo-American Media in the World* (New York: Columbia University Press, 1977); Alan Wells, *Picture Tube Imperialism: A Case Study of Television* (Maryknoll, N.Y.: Orbis Books, 1972); and Anthony Smith, *The Geopolitics of Information: How Western Culture Dominates the World* (New York: Oxford University Press, 1980).
32. For an analysis of the shortcomings of the "Western Marxists," see George Snedeker, "Western Marxism and the Problem of History," *Psychology and Social Theory,* no. 4 (1984): 41–50.

33. An illustration of the contending points of view is found in an exchange between, on the one hand, Abercrombie, Hill, and Turner (the authors of *The Dominant Ideology Thesis*), who reject the idea that ideology is shaped in the main by material conditions, and, on the other, Therborn, who fells that the relationship between consciousness and material forces can be usefully applied. N. Abercrombie, S. Hill, and R. S. Turner, "Determinacy and Indeterminacy in the Theory of Ideology," *New Left Review,* no. 142 (November–December 1983): 55–66, and Goran Therborn, "New Questions of Subjectivity," *New Left Review,* no. 143 (January–February 1984): 97–107.
34. Williams, *Marxism and Literature,* p. 110.
35. Sherman, "'Technology vis-à-vis Institutions,'" p. 182.
36. Marx, "Preface to *A Contribution to the Critique of Political Economy,*" p. 5.
37. *The Media Crisis,* (Washington, D.C.: World Press Freedom Committee, 1982). pp. 143–147. Schiller has argued that the only "world information order" had been "an American one." Herbert I. Schiller, "Breaking the West's Media Monopoly," *The Nation,* 21 September 1985, pp. 248–251. See also Kaarle Nordenstreng, with Lauri Hannikainen, *The Mass Media Declaration of UNESCO* (Norwood, N. J.: Ablex, 1984), and Oliver Boyd-Barrett, *The International News Agencies* (Beverly Hills, Calif.: Sage, 1980).
38. "First Report of the FCC Advisory Committee Made in Preparation for the ITU's WARC-ORB-85 (World Administrative Radio Conference on the Use of the Geostationary Satellite Orbit and the Planning of the Space Services Utilizing It)," December 1983, pp. 4–15.
39. "Come in, Afghanistan," *New York Times,* 1 October 1985, p. A24.
40. Quoted in Central Intelligence Agency, Joint Publication Research Service, *Worldwide Report: Telecommunications Policy, Research, and Development,* no. 250, cited in *Review of International Broadcasting,* no. 76 (May 1983): 37.
41. Karl Marx, *The Eighteenth Brumaire of Louis Bonaparte* (New York: International Publishers, 1972), p. 123.
42. Ibid., pp. 123–124.
43. Karl Marx, "A Contribution to the Critique of Political Economy," in *Communication and Class Struggle,* ed. Mattelart and Siegelaub, p. 73.
44. Victor E. Neuberg, *Popular Literature: A History and Guide* (New York: Penguin Books, 1977). Low-cost newspapers (the penny press) grew rapidly in England after repeal of longstanding legal restraints, called "taxes on knowledge" by their foes.
45. Hans Magnus Enzensberger, *The Consciousness Industry: On Literature, Politics and the Media* (New York: Seabury Press, 1974).
46. Karl Marx and Friedrich Engels, *The German Ideology* (New York: International Publishers, 1970), p. 64.
47. Antonio Gramsci, *Selections from the Prison Notebooks,* ed. Quinton Hoare and Geffrey Nowell Smith (New York: International Publishers, 1971), p. 12.
48. Ibid.
49. Pietilä, "On the Scientific Status and Position of Communication Research," p. 78.
50. G. A. Williams, "Gramsci's Concept of Egemonia," *Journal of the History of Ideas* 21 (October–December 1960): 586–599, quoted in Ralph Miliband, *The State in Capitalist Society* (New York: Basic Books, 1969), p. 180.
51. Timothy Green, *The Universal Eye* (London: Bodley Head, 1972), cited in Ju-

dith Todd, "The News Business: Bourgeoise Research and the Mass Media," *World Marxist Review* 16 (July 1973): 129–130.

52. Acosta, "Mass Media and Imperialist Ideology," p. 150.
53. Gaye Tuchman, ed., *The TV Establishment: Programming for Power and Profit.* (Englewood Cliffs, N.J.: Prentice-Hall, 1974), p. 39.
54. David L. Sallach, "Class Domination and Ideological Hegemony," in ibid., p. 167.
55. Ibid, p. 166.
56. Adapted from S. Hall, "Culture, the Media and the 'Ideological Effect,'" in *Mass Communication and Society,* ed. J. Curran, M. Gurewitch, and J. Woolacott (London: Arnold, 1977), in Raquel Salinas and Leena Paldan, "Culture in the Process of Dependent Development: Theoretical Perspectives," in *National Sovereignity and International Communication,* ed. Kaarle Nordenstreng and Herbert I. Schiller (Norwood, N.J.: Ablex, 1979), p. 85.
57. Jerry Rubin, quoted in Schiller, *The Mind Managers,* p. 150.
58. Quoted in Kathy Ferguson, "Feminism and Bureaucratic Discourse," *New Political Science,* no. 11 (Spring 1983): 53.
59. Edward W. Ploman, "Satellite Broadcasting, National Sovereignty, and Free Flow of Information," in *National Sovereignty and International Communication,* ed. Nordenstreng and Schiller, p. 162.
60. "Free flow" is only one of the slogans that have become weapons in the information and radio wars. The United States Advisory Commission on Public Diplomacy calls the problem "semantic corruption" and decries "the misuse by communist propagandists of such positive words as 'peace,' 'democratic,' 'liberation' and 'people,' to mention only the most obvious." The commission reported that the United States Information Agency had prepared a "lexicon of Marxist-Leninist terminology" and sent it to all overseas posts so that personnel could be forewarned. *United States Advisory Commission on Public Diplomacy–1985 Report* (Washington, D.C.: USACPD, 1985), p. 16.
61. Acosta, "Mass Media and Imperialist Ideology," pp. 149–150.
62. Armand Mattelart, "Communication Ideology and Class Practice," in *Communication and Class Struggle,* ed. Mattelart and Siegelaub, p. 115.
63. Ibid.
64. Herbert I. Schiller, *Mass Communication and American Empire* (New York: Kelley, 1969); *The Mind Managers; Communication and Cultural Domination; Who Knows;* and, with Nordenstreng, *National Sovereignty and International Communication.*
65. Tapio Varis, *Television Traffic: A One-Way Street,* UNESCO Series on Mass Communications, no. 71 (New York: UNESCO, 1972).
66. Wells, *Picture Tube Imperialism.*
67. Richard Bunce, *Television in the Corporate Interest* (New York: Praeger, 1976).
68. Tunstall, *The Media Are American.*
69. Smith, *The Geopolitics of Information.*
70. S. B. O. Gutto, "Captivity of the Press," *Monthly Review* 34 (January 1983): 31–41.
71. Two useful explications of how Hegel perceived the master–slave relationship in *Phenomenology of Mind* are in Charles Taylor, *Hegel* (Cambridge: Cambridge University Press, 1975), pp. 152–157, and Frederick Copleston, *A History of Philosophy,* (Garden City, N.Y.: Doubleday, 1963), vol. 7, pp. 221–225.

72. Mosco and Herman, "Radical Social Theory and the Communications Revolution," p. 3.
73. Fredric Jameson, "Postmodernism, or the Cultural Logic of Late Capitalism," *New Left Review*, no. 146 (July–August, 1984): 53–92. Borrowing from Ernest Mandel, Jameson posited three successive stages of capitalism: (1) market, (2) monopoly, and (3) multinational (or "late" or "consumer"). Each corresponds to a specific stage of the technological revolution: (1) since 1848, machine production of machines (steam-driven motors); (2) since the 1890s, machine production of electric and combustion motors; and (3) since the 1940s, machine production of electronic and nuclear-powered apparatuses (p. 78).
74. Mosco and Herman, "Radical Social Theory and the Communications Revolution," p. 6.
75. Quoted in Armand Mattelart, "Mass Communications in the Third World: Theory and Practice," *Latin American Perspectives* 5 (Winter 1978): 14.
76. Quoted in Schiller, *Mass Communication and American Empire*, p. 7. In a speech given in 1973, Sarnoff estimated that the satisfactory resolution of the controversy over communications satellites—which he envisaged to be "the commitment of all nations to the eventual goal of unrestricted programming"—would offer "rich rewards," including "billions of dollars in fresh opportunities for international trade." Robert W. Sarnoff, "Freedom to Communicate" (Speech, February 9, 1973).
77. Gramsci, *Selections from the Prison Notebooks*, p. 54.
78. "Let the public not only be educated, but let it be itself the educator." Bertolt Brecht, *Sur le Cinema* (Paris: l'Arche, 1970), p. 139.
79. "But it is not at all our task to restore these ideological institutions by innovations that build on the existing social order. We ought rather to use our innovations to shake their foundation: up with innovation, down with restoration!" Ibid., p. 141.
80. Frantz Fanon, *A Dying Colonialism* (New York: Grove Press, 1965), pp. 69–97.
81. Ibid., p. 69.
82. Ibid., p. 76.
83. Ibid., p. 84.
84. For an account of the clandestine radio of the African National Congress, see Alexander Sibeko, "The Underground Voice," in *Communication and Class Struggle*, ed. Armand Mattelart and Seth Siegelaub (New York: International General, 1983), vol. 2, pp. 203–207.
85. Julian Hale, *Radio Power: Propaganda and International Broadcasting* (Philadelphia: Temple University Press, 1975), p. 104.
86. Hans Magnus Enzensberger, "Constituents of a Theory of the Media," *New Left Review*, no. 64 (November–December 1970): 15.
87. Schiller, *Communication and Cultural Domination*, p. 68.
88. Ibid., p. 76.
89. Salinas and Paldán, "Culture in the Process of Dependent Development," p. 93.
90. Karl Marx, *Capital* (New York: Pelican, 1976), vol. 1, p. 1048. A striking example of this process was illustrated on the "MacNeil/Lehrer NewsHour" on October 1, 1985 as Charlayne Hunter-Gault interviewed a black African advertising executive who worked for a large, presumably U.S.-based, agency

in Johannesburg and who went home each night to the "lived reality" of Soweto.

91. Salinas and Paldán, "Culture in the Process of Dependent Development," p. 93.
92. Ibid., p. 97.

Chapter 7

1. Japan has been in the forefront of "high-definition" television technology, which involves a digital system of signal projection and processing. The "high definition" is possible as a result of transmitting signals via digital code—the "native tongue of the computer"—rather than via the analog system, which is a continuous electrical signal. The television screen, in effect, becomes the display component of a minicomputer. For a popular account of "high-definition" television, see Igor Rosselevich, "Digital TV: A Revolution in Mass Media," *Soviet Life,* October 1985, pp. 25–26.
2. Olof Lundberg (Director General of INMARSAT [International Maritime Satellite Organization], London), "Mobile Communications in the World Market," *Telephony,* 29 July 1985; reprinted in *Teleclippings* [ITU-Geneva], No. 781, 26 August 1985, p. 6.
3. The Soviet communications satellites that are being tracked do not use the geostationary orbit, but circle the earth in an elliptical orbit designed to accommodate the northern land mass and eleven time zones of that country. This orbit puts them in reach of North American receiving stations.
4. One amateur expert enthusiastically reports of having tuned in via his 10-foot dish antenna to the audio (radio) signals emanating from satellite transponders (these signals are the nonvideo services of communications satellites). Classical music came across with "pristine clarity . . . all of which makes me long even more for International Radio by satellite as an alternative or supplement to High Frequency broadcasting." Loren Cox, Jr., "Radio Equipment Forum," *Review of International Broadcasting,* no. 104 (1985): 33.
5. Jonathan Friendly, "U.S. Is Studying Effects of Antennas' Radiation," *New York Times,* 24 July 1985, p. B2.
6. While a merger of the two systems is unlikely, there has been a measure of ongoing cooperation. Both were accorded "permanent observer status" at the 1985 session of COPUOS and have been given nonvoting participant status in the ITU. However in late 1986, the head of Intelsat, Richard R. Colino, was ousted. His replacement, Dean Burch, was supported by the Reagan administration.
7. The obvious contradiction between the policies based on the ideal of "free flow of information" and the restrictions on the flow of scientific information is noted here only in passing. Indeed, the whole question of "flow" is fraught with contradiction. Consider how, at the same time that the United States government cuts off technical and scientific information flows, former bank president and State Department adviser Walter B. Wriston decries the rise of barriers set by *other* governments against data-processing facilities and outgoing transfers of information, which, he argues, they wrongly seek to tax or regulate. Walter B. Wriston, "Policy Formation for the Global Marketplace," *Chronicle of International Communication,* 6 (January–February 1985): 1, 7–8.

8. *Wall Street Journal,* cited in *The Interdependent* 2 (July–August 1985): 1.
9. Charles J. Rolo, *Radio Goes to War: The "Fourth Front"* (New York: 1942), p. 244. At present, five nongovernment shortwave stations broadcast internationally from the United States, all of which carry religious programs. The FCC has authorized international broadcasting by a commercial shortwave station, NDXE (pronounced "In Dixie"), with transmitters located in Opelika, Alabama. It has announced that it will carry advertising by multinationals like Coca-Cola as well as tobacco companies, and will give its global listeners the opportunity to order products from the United States. Rhonda Johnston, "New Radio Station Broadcasts to World from Dixie," *The Auburn* (Alabama) *Bulletin,* 13 February 1985, p. A-2.
10. As of late 1986, these broadcasts were reported to have been abandoned.
11. One possibility would be to distribute free equipment in targeted areas, as was done during the Vietnam War with transistor radios, in order to inform and persuade the population.
12. A hint as to the complexity of getting locations for transmitters comes from a report heard on Radio KOL (Israel) on September 1, 1985. It stated that the Israeli government had decided on three conditions for approving the operation of a Voice of America relay: (1) no anti-Soviet content; (2) Asiatic beam—no transmissions to the European part of the Soviet Union; and (3) no broadcasts in Yiddish or Hebrew. *Review of International Broadcasting,* no. 102, p. 18.
13. *United States Advisory Commission on Public Diplomacy–1985 Report* (Washington, D.C.: USACPD, 1985), pp. 17–19.
14. Ibid., p. 17.
15. Quoted in "MIT Study Defines Dilemma," *Chronicle of International Communication* 5 (September, 1984): 16.
16. Leslie Milk and Allen Weinstein, "U.S. Participation in the International Telecommunications Union: A Study of Policy Alternatives" (Washington, D.C.: Center for Strategic and International Studies, Georgetown University, 1984, Mimeographed).
17. William J. Broad, "Secrecy to Surround Atlantis's First Flight," *New York Times,* 1 October 1985, pp. C1, C7.
18. Eric Pace, "TV Curb Is Backed by U.N. Assembly," *New York Times,* 11 December 1982, p. 6.
19. Ibid.
20. The United Nations Association of the United States of America's 1985 annual survey of issues before the United Nations comments: "The U.S., a major sensing state, has appeared reluctant to sacrifice profit, especially for the private companies that own the satellites or analyse the data." Frederic Eckhard, ed., *Issues 40: Issues Before the 40th General Assembly of the United Nations* (New York: UN/USA, 1985), p. 142.
21. "Communications Development Issues Occupy Ad Hoc Policy Group," *Chronicle of International Broadcasting* 6 (May 1985): 1, 7–8.
22. See R. G. Amero, R. R. Bowen, J. G. Chambers, and J. Forsey, "The Results of RARC-83: Broadcasting Satellite Planning for Region 2," *Space Communication and Broadcasting* 1 (December 1983): 339–349.
23. U.N. General Assembly, Committee on the Peaceful Uses of Outer Space, *The Feasibility of Using Direct Broadcast Satellites for Educational Purposes and of Internationally or Regionally Owned Space Segments,* A/AC.105/341/Rev.1.29, 1985.

24. At present, the United Nations rents space on the Voice of America's transmitters to issue its public-affairs shortwave broadcasts, an arrangement that has not gone unchallenged in the current anti-United Nations atmosphere: "America's best and most powerful shortwave station is thus handed over as a mouthpiece to the national policies of more than 100 countries, most of which regularly throw stones at us from the glass house beside the East River." Leigh Lerner, "UN Radio: U.S.–Subsidized Lies and Rhetoric," *Minneapolis Star and Tribune,* 19 April 1985, p. 31. Distribution of the United Nations broadcasts is very limited within the United States. To my knowledge, WBAI/Pacifica Radio is the only station in the New York tri-state area that carries them regularly.
25. U.N. General Assembly, Committee on the Peaceful Uses of Outer Space, *Coordination of Outer Space Activities Within the United Nations System: Programme of Work for 1985 and 1986 and Future Years.* Report of the Secretary General, 1984.

Selected Bibliography

This bibliography has been broken down into the following headings:

- International Communications History
- Mass Communication, Ideologies, and Development
- Early Studies of Mass Communication
- Mass Communication Sociology in the United States Since World War II
- International Law and International Broadcasting
  - Studies Published Before 1970
  - Studies Published Since 1970
- Critical and Marxist Approaches to Mass Communications
- Outer Space, International Broadcasting, and the Direct Broadcast Satellite (DBS)

International Communications History

Aitken, Hugh, G. J. *The Continuous Wave: Technology and American Radio, 1900–1932*. Princeton, N.J.: Princeton University Press, 1985.

———. *Syntony and Spark: The Origins of Radio*. New York: Wiley, 1975; Princeton, N.J.: Princeton University Press, 1985.

Archer, Gleason L. *Big Business and Radio*. New York: American Historical Company, 1939.

———. *History of Radio to 1926*. New York: American Historical Company, 1938.

Baker, Ray Stannard. *Woodrow Wilson and World Settlement*. 3 vols. Garden City, N.Y.: Doubleday, 1922.

Beard, Charles A., and William Beard. *The American Leviathan: The Republic in the Machine Age*. New York: Macmillan, 1930.

Becker, Carl L. *History of Modern Europe*. Washington, D.C.: Silver Burdett, 1945.

Borchardt, Kurt. *Structure and Performance of the U.S. Communications Industry: Government Regulation and Company Planning*. Boston: Harvard Graduate School of Business Administration, 1970.

Bozhenkova, Margarita. "Alexander Popov in Chicago." *Soviet Life*, October 1985, pp. 27–28.

Clark, Keith. *International Communications, the American Attitude*. New York: Columbia University Press, 1931.

Clarke, Arthur C. *The Exploration of Space*. New York: Pocket Books, 1951.

———. *The Making of a Moon: The Story of the Earth Satellite Program*. New York: Harper, 1957.

———. *Voices Across the Sea*. New York: Harper & Row, 1974.

Codding, George A., Jr. *The International Telecommunication Union, an Experiment in International Cooperation*. London: Brill, 1952.

Codding, George A. Jr., and Anthony M. Rutkowski. *The International Telecommunication Union in a Changing World*. Dedham, Mass.: Artech House, 1982.

Duboff, Richard B. "The Rise of Communications Regulation: The Telegraph Industry, 1844–1880." *Journal of Communications* 34 (Summer 1984): 52–66.

Edelman, Jacob Murray. *The Licensing of Radio Services in the United States, 1927 to 1947: A Study in Administrative Formulation of Policy*. Urbana: University of Illinois Press, 1950.

Emery, Walter B. *National and International Systems of Broadcasting: Their History, Operation, and Control*. East Lansing: Michigan State University Press, 1969.

Encyclopaedia Britannica. 11th ed. 29 vols. New York: Cambridge University Press, 1911.

Giraud, Chester, and Garnet R. Garrison. *Radio and Television*. New York: Appleton-Century-Crofts, 1950.

Grandin, Thomas. *The Political Use of the Radio*. Geneva: Geneva Research Center, League of Nations, 1939.

Herring, James M., and Gerald C. Gross. *Telecommunications: Economics and Regulation*. New York: McGraw-Hill, 1936.

Hoover, Herbert. *The Memoirs of Herbert Hoover: The Cabinet and the Presidency, 1920–1933*. New York: Macmillan, 1952.

International Telecommunication Conferences. Report of the U.S. delegation to the International Telecommunication Union conference. Washington, D.C.: Government Printing Office, 1948.

Larkin, Oliver W. *Samuel F. B. Morse and American Democratic Art*. Boston: Little, Brown, 1954.

Long, Stewart Lovis. *The Development of the Television Network Oligopoly*. New York: Arno Press, 1979.

Mabee, Carleton, *The American Leonardo: A Life of Samuel F. B. Morse*. New York: Knopf, 1943.

Marconi, G. "Where Is Radio Going?" *Saturday Evening Post,* 3 December 1927, pp. 8–9, 44–48.

Martin, James. *Telecommunications and the Computer*. Englewood Cliffs, N.J.: Prentice-Hall, 1969.

Mazlish, Bruce, ed. *The Railroad and the Space Program: An Exploration in Historical Analogy*. Cambrige, Mass.: MIT Press, 1965.

New York Times. Coverage of and commentary on international radio conferences held in 1903 to 1906, 1912, 1925, 1927, and 1932.

Rolo, Charles J. *Radio Goes to War: The "Fourth Front."* New York: Putnam, 1942.

Sarnoff, David. "U.S. Counterattack." In *A Guide to Anti-Communist Action*. Edited by A. T. Bouscaren. Chicago: Regnery, 1958.

Tomlinson, John D. *The International Control of Radio Communications*. Ann Arbor, Mich.: Edwards, 1945.

Tribolet, Leslie Bennett. *The International Aspects of Electrical Communications in the Pacific Area*. Baltimore: Johns Hopkins University Press, 1929.

U.S. Naval War College. *International Law Situations, with Solutions and Notes*. Washington, D.C.: Government Printing Office, 1907.

Waldrop, Frank C., and Joseph Borkin. *Television: A Struggle for Power*. New York: Morrow, 1938; New York: Arno Press, 1971.

Ware, Norman. *The Industrial Worker, 1840–60: The Reaction of American Industrial Society to the Advance of the Industrial Revolution*. New York: Quadrangle Press, 1964.

White, Llewellyn, and Robert D. Leigh. *Peoples Speaking to Peoples: A Report on International Mass Communication for the Commission on Freedom of the Press*. Chicago: University of Chicago Press, 1946.

Young, Owen D. "Freedom of the Air." *Saturday Evening Post*, 16 November 1929, pp. 16, 193–194.

Mass Communication, Ideologies, and Development

Agee, Philip. *Inside the Company: CIA Diary*. New York: Bantam Books, 1975.

Althusser, Louis. *Philosophy and Other Essays*. New York: Monthly Review Press, 1971.

Ambrose, Stephen E. *Rise to Globalism: American Foreign Policy Since 1938*. Baltimore: Penguin, 1971.

Amin, Samir. *Imperialism and Unequal Development*. New York: Monthly Review Press, 1977.

Amin, Samir, G. Arrighi, and Andre Gunder Frank. *Dynamics of Global Crisis*. New York: Monthly Review Press, 1982.

Avineri, Shlomo, ed. *Karl Marx on Colonialism and Modernization*. New York: Doubleday, Anchor Books, 1969.

Baran, Paul A. *The Political Economy of Growth*. New York: Monthly Review Press, 1957.

Baran, Paul A., and Paul M. Sweezy. *Monopoly Capital*. New York: Monthly Review Press, 1966.

Barnet, Richard J. *The Alliance: America–Europe–Japan, Makers of the Postwar World*. New York: Simon and Schuster, 1983.

———. *Intervention and Revolution: America's Confrontation with Insurgent Movements Around the World*. New York: World, 1968.

Barnet, Richard J., and Ronald E. Muller. *Global Reach: The Power of the Multinational Corporations*. New York: Simon and Schuster, 1974.

Becker, Jörg, Göran Hedebro, and Leena Paldán, eds. *Communication and Domination: Essays to Honor Herbert I. Schiller*. Norwood, N.J.: Ablex, 1986.

Beckford, G. L. *Persistent Poverty: Underdevelopment in Plantation Economics of the Third World*. New York: Oxford University Press, 1972.

Benn, Stanley I. "State." *Encyclopedia of Philosophy*. 8 vols. New York: Macmillan and The Free Press, 1967.

Berlin, Isaiah. "Nationalism: Past Neglect and Present Power." *Partisan Review* 46, no. 3 (1979): 337–358.

Berman, Edward H. *The Ideology of Philanthropy: The Influence of the Carnegie, Ford, and Rockefeller Foundations on American Foreign Policy*. Albany: State University of New York Press, 1983.

Bosch, Juan. *Pentagonism: A Substitute for Imperialism*. New York: Grove Press, 1968.

Boulding, Kenneth E. *Economic Imperialism*. Ann Arbor: University of Michigan Press, 1972.

Carr, Edward Hallett. *The New Society*. Boston: Beacon Press, 1951.

Chaliand, Gerard. *Revolution in the Third World*. New York: Penguin, 1977.

Chasin, Barbara, and Gerald Chasin. *Power and Ideology: A Marxist Approach to Political Sociology*. Cambridge, Mass.: Schenkman, 1974.

Chinweizu. *The West and the Rest of Us: White Predators, Black Slavers, and the African Elite*. New York: Random House, 1975.

CIA, the Press, and Central America. Special issue. *Covert Action, Information Bulletin*, no. 21 (Spring 1984).

Cockcroft, James D., Andre Gunder Frank, and Dale L. Johnson. *Dependency and Underdevelopment: Latin America's Political Economy*. New York: Doubleday, Anchor Books, 1972.

Colletti, Lucio. *From Rousseau to Lenin: Studies in Ideology and Society*. New York: Monthly Review Press, 1972.

Crozier, Michael J., Samuel P. Huntington, and Joji Watanuki. *The Crisis of Democracy: Report on the Governability of Democracies to the Trilateral Commission*. New York: New York University Press, 1975.

Deutsch, Karl W. *Nationalism and Social Communication: An Inquiry into the Foundation of Nationality*. Cambridge, Mass.: MIT Press, 1953, 1969.

Dixon, Marlene, and Suzanne Jonas. *Revolution and Intervention in Central America*. San Francisco: Synthesis Press, 1979.

Dobb, Maurice. *Studies in the Development of Capitalism*. New York: International Publishers, 1947.

Dowd, Douglas. *Modern Economic Problems in Historical Perspective*. Boston: Heath, 1966.

Draper, Hal. *Karl Marx's Theory of Revolution*. 2 vols. New York: Monthly Review Press, 1977.

Dumont, René, with Marcel Mazoyer. *False Start in Africa*. New York: Praeger, 1973.

Emmanuel, Arghiri. *Unequal Exchange: A Study of the Imperalisms of Trade*. With comments by Charles Bettelheim. New York: Monthly Review Press, 1972.

Erlich, Alexander. "Neo-Marxian and Neo-neo-Marxian Theories of Imperialism." Paper presented at Vassar College seminar, Poughkeepsie, New York, April 5, 1977.

Facing the 1980s: New Directions in the Theory of Imperialism. Special issue. *Review of Radical Political Economics* 11 (Winter 1979).

Fann, K. T., and Donald C. Hodges, eds. *Readings in U.S. Imperialism*. Boston: Porter Sargent, 1971.

Fanon, Frantz. *A Dying Colonialism*. New York: Grove Press, 1965.

———. *Toward the African Revolution*. New York: Grove Press, 1967.

———. *The Wretched of the Earth*. New York: Grove Press, 1963.

Finger, Seymour Maxwell. "United States Policy Toward International Institutions." *International Organization* 30 (Spring 1976): 347–360.

Frank, Andre Gunder. "Disarray in the Atlantic Alliance." *Our Socialism: A Journal by American Socialists* 1 (March 1983): 27–44.

———. *Latin America: Underdevelopment or Revolution*. New York: Monthly Review Press, 1969.

———. *World Accumulation, 1492–1789*. New York: Monthly Review Press, 1978.

Franklin, Raymond S. *American Capitalism: Two Visions*. New York: Random House, 1977.

Fulbright, J. William. *The Pentagon Propaganda Machine*. New York: Liveright, 1970.

Gilbert, Tony, and Pierre Joris. *Global Interference: The Consistent Pattern of American Foreign Policy*. London: Liberation Press, 1981.

Gramsci, Antonio. *Selections from the Prison Notebooks*. Edited by Quinton Hoare and Geffrey Nowell Smith. New York: International Publishers, 1971.

Greene, Felix. *The Enemy: What Every American Should Know About Imperialism*. New York: Random House, Vintage Books, 1970.

Gudykunst, William B., ed. *Intercultural Communications: Current Perspectives*. Beverly Hills, Calif.: Sage, 1983.

Hedebro, Goran. *Communication and Social Change in Developing Nations: A Critical View*. Ames: Iowa State University Press, 1982.

Heilbroner, Robert. *The Limits of American Capitalism*. New York: Harper & Row, 1966.

Hoffman, Stanley. *Primacy or World Order: American Foreign Policy Since the Cold War*. New York: McGraw-Hill, 1978.

Hopkins, Terence K., Immanuel Wallerstein et al. *World-Systems Analysis: Theory and Methodology*. Beverly Hills, Calif.: Sage, 1982.

Horowitz, David. *The Free World Colossus: A Critique of American Foreign Policy in the Cold War*. New York: Hill and Wang, 1965.

Horowitz, Irving Louis. "Capitalism, Communism, and Multinationalism." *Society* 11 (January–February 1974): 32–43.

Huberman, Leo. *Man's Worldly Goods: The Story of the Wealth of Nations*. New York: Monthly Review Press, 1959.

Jalee, Pierre. *How Capitalism Works*. New York: Monthly Review Press, 1977.

Jones, Peter d'A. *The Consumer Society. A History of American Capitalism*. Baltimore: Penguin, 1965.

Kautsky, John J., ed. *Political Change in Underdeveloped Countries: Nationalism and Communism*. New York: Wiley, 1967.

Lenin, V. I. *Imperialism, the Highest Stage of Capitalism*. New York: International Publishers, 1939.

———. *The State and Revolution*. New York: International Publishers, 1932.

Lerner, Daniel. "Notes on Communication and the Nation-State." *Public Opinion Quarterly* 37 (Winter 1973/74): 541–550.

Lichtheim, George. *The Concept of Ideology and Other Essays*. New York: Random House, 1967.

———. *Imperialism*. New York: Praeger, 1971.

Luxemburg, Rosa. *The Accumulation of Capital—an Anti-Critique*. Published with Bukharin, Nikolai I. *Imperialism and the Accumulation of Capital*. New York: Monthly Review Press, 1972.

Magdoff, Harry. *The Age of Imperialism: The Economics of U.S. Foreign Policy*. New York: Monthly Review Press, 1969.

Marchetti, Victor, and John D. Marks. *The CIA and the Cult of Intelligence*. New York: Dell, 1975.

Marx, Karl. *Capital*. New York: International Publishers, 1967.

———. *Critique of the Gotha Programme*. Moscow: Progress Publishers, 1971.

———. *Economic and Philosophical Manuscripts*. New York: International Publishers, 1964.

———. *The Eighteenth Brumaire of Louis Bonaparte*. New York: International Publishers, 1972.

———. *Grundrisse: Foundations of the Critique of Political Economy*. New York: Random House, Vintage Books, 1973.

———. *Selected Writings in Sociology and Social Philosophy*. Edited by T. B. Bottomore. New York: McGraw-Hill, 1961.

Marx, Karl, and Friedrich Engels. *The German Ideology*. Part 1. New York: International Publishers, 1970.

Memmi, Albert. *The Colonizer and the Colonized*. Boston: Beacon Press, 1965.

Miliband, Ralph. *The State in Capitalist Society*. New York: Basic Books, 1969.

Mintz, Morton, and Jerry S. Cohen. *Power, Inc.: Who Really Runs America?* New York: Bantam Books, 1977.

Mommsen, Wolfgang. *Theories of Imperialism*. Chicago: University of Chicago Press, 1980.

Mydral, Gunnar. *Asian Drama: An Inquiry into the Poverty of Nations*. New York: Random House, 1971.

Nabudere, Dan. *The Political Economy of Imperialism*. London: Zed Press, 1977.

Nearing, Scott, and J. Freeman. *Dollar Diplomacy: A Study in American Imperialism*. New York: Huebsch and The Viking Press, 1925.

O'Connor, James. *The Corporations and the State: Essays in the Theory of Capitalism and Imperialism*. New York: Harper & Row, 1974.

Offe, Claus. "Political Authority and Class Structures." In *Critical Sociology*. Edited by Paul Connerton. New York: Penguin, 1976.

Petras, James, and Robert Rhodes. "The Reconsolidation of U.S. Hegemony." *New Left Review* (May–June 1976): 37–53.

Poulantzos, Nicos. *Classes in Contemporary Capitalism*. London: Verso, 1978.

———. "Internationalization of Capitalist Relations and the Nation-State." *Economy and Society* 3 (May 1974): 145–179.

———. *Political Power and Social Classes*. London: New Left Books, 1975.

Rhodes, R. I., ed. *Imperialism and Underdevelopment*. New York: Monthly Review Press, 1970.

Rostow, W. W. *The Stages of Economic Growth: A Non-Communist Manifesto*. New York: Cambridge University Press, 1958, 1963.

Schumpeter, Joseph. *Imperialism*. New York: New American Library, Signet, 1951.

Shafer, Boyd C. *Faces of Nationalism*. New York: Harcourt Brace Jovanovich, 1972.

Shoup, Laurence H., and William Minter. *Imperial Brain Trust: The Council on Foreign Relations and United States Foreign Policy*. New York: Monthly Review Press, 1977.

Sigmund, Paul E., Jr., ed. *The Ideologies of the Developing Nations*. New York: Praeger, 1963.

Skocpol, Theda. *States and Social Revolutions: A Comparative Analysis of France, Russia, and China*. London: Cambridge University Press, 1979.

Staley, Eugene. *The Future of Underdeveloped Countries*. New York: Harper & Row, for the Council on Foreign Relations, 1954.

Stephenson, Hugh. *The Coming Clash: The Impact of Multinational Corporations on National States*. New York: Saturday Review Press, 1972.

Sutcliffe, Robert B., and Edward Roger John Owen. *Studies in the Theory of Imperialism*. London: Longman, 1972.

Szymanski, Albert. *The Capitalist State and the Politics of Class*. Cambridge, Mass: Winthrop, 1978.

Trilateral Commission. *Seeking a New Accommodation in World Commodity Markets*. The Triangle Papers, no. 10. New York: Trilateral Commission, 1976.

Vasquez, John A. *The Power of Power Politics*. New Brunswick, N.J.: Rutgers University Press, 1983.

Wallerstein, Immanuel. "The Rise and Future Demise of the World Capitalist System: Concepts for Comparative Analysis." *Comparative Studies in Society and History* vol. 16, no. 4 (1974): 387–415.

Ward, Barbara. *Spaceship Earth*. New York: Columbia University Press, 1966.

Warren, Bill. "Imperialism and Capitalist Industrialization: Myths of Underdevelopment." *New Left Review*, no. 81 (September–October 1973): 3–44.
Wenger, Martha. "USIA/CIA: Arrowhead of Penetration." *Counterspy* 5 (November 1980–January 1981): 5–13.
Wriston, Walter. "Policy Formation for the Global Marketplace." *Chronicle of International Communications*. 6 (January–February 1985): 1, 7–8.
Yarmolinsky, Adam. *The Military Establishment*. New York: Harper & Row, 1970.
Zeitlin, Irving M. *Capitalism and Imperialism: An Introduction to Neo-Marxian Concepts*. Chicago: Markham, 1972.

Early Studies of Mass Communication

Arnheim, Rudolf. *Radio*. New York: Arno Press, 1971; New York: Da Capo Press, 1972.
Blumer, Herbert. "Elementary Collective Behavior." In *New Outline of the Principles of Sociology*. Edited by Alfred McClung Lee. New York: Barnes and Noble, 1939.
Cantril, Hadley. *The Invasion from Mars: A Study in the Psychology of Panic*. Princeton, N.J.: Princeton University Press, 1940.
Cantril, Hadley, and Gordon W. Allport. *The Psychology of Radio*. New York: Harper & Brothers, 1935.
Childs, Harwood Lawrence, and John B. Whitton, eds. *Propaganda by Short Wave*. Princeton, N.J.: Princeton University Press, 1942.
Clark, George Norman. *Unifying the World*. London: Swarthmore Press; New York: Harcourt Brace and Howe, 1920.
Lasswell, Harold D. *Democracy Through Public Opinion*. Menasha, Wis.: Banta, 1941.
———. *Politics: Who Gets What, When, How*. New York: McGraw-Hill, 1936.
———. *Propaganda Technique in the World War*. New York: Knopf, 1927.
———. *World Politics and Personal Insecurity*. New York: McGraw-Hill, 1935.
Lasswell, Harold D., and Dorothy Blumenstock. *World Revolutionary Propaganda: A Chicago Study*. New York: Knopf, 1939.
Lasswell, Harold D., Ralph D. Casey, and Bruce Lannes Smith. *Propaganda and Promotional Activities: An Annotated Bibliography*. Minneapolis: University of Minnesota Press, 1935; Chicago: University of Chicago Press, 1969.
Lazarsfeld, Paul F. "An Episode in the History of Social Research: A Memoir." In *The Intellectual Migration: Europe and America, 1930–1960*. Edited by Donald H. Fleming and Bernard Bailyn. Cambridge, Mass.: Harvard University Press, Belknap Press, 1969.
———. *Radio and the Printed Page: An Introduction to the Study of Radio and its Role in the Communication of Ideas*. New York: Duell, Sloan, and Pearce, 1940; New York: Arno Press, 1972.
Lee, Alfred McClung, and Elizabeth Briant Lee. *The Fine Art of Propaganda*. New York: Harcourt, Brace, for the Institute for Propaganda Analysis, 1939.
Lippmann, Walter. "The World Outside and the Pictures in Our Heads." In *Images of Man: The Classic Tradition in Sociological Thinking*. Edited by C. Wright Mills. New York: Braziller, 1960.
Mills, C. Wright, ed. *Images of Man: The Classic Tradition in Sociological Thinking*. New York: Braziller, 1960.

Riegel, Oscar W. *Mobilizing for Chaos: The Story of the New Propaganda*. New Haven, Conn.: Yale University Press, 1934.
Rogerson, Sidney. *Propaganda in the Next War*. London: Geoffrey Books, 1938.
Smith, Bruce L., Harold D. Lasswell, and Ralph D. Casey. *Propaganda, Communication, and Public Opinion: A Comprehensive Reference Guide*. Princeton, N.J.: Princeton University Press, 1946.
Waller, Judith C. *Radio: The Fifth Estate*. Boston: Houghton Mifflin, 1946.
Wilke, Walter Herman. *An Experimental Comparison of the Speech, the Radio, and the Printed Page as Propaganda Devices*. New York, 1934. Issued also as Ph.D. thesis, Columbia University.
Willey, Malcolm Macdonald, and Stuart A. Rice. *Communications Agencies and Social Life*. New York: McGraw-Hill, 1933.

Mass Communication Sociology in the United States Since World War II

(Includes theoretical and empirical studies of mass media in the United States, works dealing with international communication, and with communications in the Soviet Union and in Third World countries.)

Aspen Institute for Humanistic Studies. *The Social Impact of Television: A Research Agenda for the 1980's*. New York: AIHS, 1981.
Barnouw, Erik. *The Image Empire*. New York: Oxford University Press, 1970.
———. *The Sponsor: Notes on a Modern Potentate*. New York: Oxford University Press, 1978.
———. *Tube of Plenty*. New York: Oxford University Press, 1975.
Berelson, Bernard. *Content Analysis in Communication Research*. Glencoe, Ill.: Free Press, 1952.
Berelson, Bernard, and Morris Janowitz, eds. *Reader in Public Opinion and Communication*. Glencoe, Ill.: Free Press, 1950.
Brown, J. A. C. *Techniques of Persuasion: From Propaganda to Brainwashing*. Baltimore: Penguin, 1963.
Bryson, Lyman, ed. *The Communication of Ideas, a Series of Addresses*. New York: Harper and Brothers, for the Institute for Religious and Social Studies, 1948.
Cantril, Hadley. *Soviet Leaders and Mastery over Man*. New Brunswick, N.J.: Rutgers University Press, 1960.
Chafee, Zechariah, Jr. *Government and Mass Communications*. A Report for the Commission on Freedom of the Press. 2 vols. Chicago: University of Chicago Press, 1947.
Cherry, Colin. *On Human Communication*. Boston: MIT Press, 1957.
———. *World Communication: Threat or Promise? A Socio-technical Approach*. New York: Wiley, 1971, 1978.
Chester, Giraud, and Garnet R. Garrison. *Radio and Television: An Introduction*. New York: Appleton-Century-Crofts, 1950.
Comstock, George. *Television in America*. Beverly Hills, Calif.: Sage, 1980.
Coombs, P. H. *The Fourth Dimension of Foreign Policy*. New York: Harper & Row, 1964.

Covert, Catherine L., and John L. Stevens, eds. *Mass Media Between the Wars: Perceptions of Cultural Tension, 1918–1941*. Syracuse, N.Y.: Syracuse University Press, 1984.

Davison, Walter Phillips. *International Political Communication*. New York: Praeger, 1965.

———. "Public Opinion." In *International Encyclopedia of the Social Sciences*. Vol. 13. Edited by David L. Sills. New York: Macmillan, 1968.

DeFleur, Melvin L. *Theories of Mass Communication*. New York: McKay, 1966.

Dewhirst, Martin, and Robert Farrell, eds. *The Soviet Censorship*. Metuchen, N.J.: Scarecrow Press, in cooperation with Radio Liberty Committee (New York) and the Institute for the Study of the Soviet Union (Munich), 1973.

Dexter, Lewis Anthony, and David Manning White, eds. *People, Society, and Mass Communications*. New York: Free Press, 1964.

Dizard, Wilson P. *The Strategy of Truth: The Story of the U.S. Information Service*. Washington, D.C.: Public Affairs Press, 1961.

Doob, Leonard W. *Communication in Africa: A Search for Boundaries*. New Haven, Conn.: Yale University Press, 1961

Dreitzel, Hans Peter. *Recent Sociology No. 2: Patterns of Communicative Behavior*. New York: Macmillan, 1970.

Edelman, Jacob Murray. *Politics as Symbolic Action: Mass Arousal and Quiescence*. Chicago: Markham, 1971.

Ellul, Jacques. *Propaganda: The Formation of Men's Attitudes*. New York: Knopf, 1965.

Etzioni, Amitai. "Nonconventional Uses of Sociology as Illustrated by Peace Research." In *The Uses of Sociology*. Edited by Paul F. Lazarsfeld, William H. Sewell, and Howard L. Wilensky. New York: Basic Books, 1967.

Ferment in the Field. Special issue on mainstream and critical approaches to the study of mass media. *Journal of Communications* 33 (Summer 1983).

Ganley, Oswald H., and Gladys D. Ganley. *To Inform or to Control: The New Communications Network*. New York: McGraw-Hill, 1982.

Gerbner, George, ed. *Mass Media Policies in Changing Cultures*. New York: Wiley, 1977.

Gerbner, George, Larry P. Gross, and William H. Melody, eds. *Communications Technology and Social Policy: Understanding the New "Cultural Revolution."* New York: Wiley, 1973.

Glazer, Nathan. "The Ideological Uses of Sociology." In *The Uses of Sociology*. Edited by Paul F. Lazarsfeld, William H. Sewell, and Howard L. Wilensky. New York: Basic Books, 1967.

Gouldner, Alvin W. *The Coming Crisis of Western Sociology*. New York: Basic Books, 1970.

Guback, Thomas H. "Film as International Business." *Journal of Communication* 24 (Winter 1974): 90–101.

Hale, Julian. *Radio Power: Propaganda and International Broadcasting*. Philadelphia: Temple University Press, 1975.

Hollander, Gayle Durham. "Political Communication and Dissent in the Soviet Union." In *Dissent in the USSR*. Edited by Rudolf L. Tokes. Baltimore: Johns Hopkins University Press, 1975.

———. *Soviet Political Indoctrination: Developments in Mass Media and Propaganda Since Stalin*. New York: Praeger. 1973.

Holt, Robert T. *Radio Free Europe*. Minneapolis: University of Minnesota Press, 1958.

Hopkins, Mark W. *Mass Media in the Soviet Union*. Pegasus, N.Y.: Western, 1970.

Inkeles, Alex. *Public Opinion in Soviet Union: A Study in Mass Persuasion*. Cambridge, Mass.: Harvard University Press, 1950.

———. *Social Change in Soviet Russia*. Russian Research Center Studies, no. 57. Cambridge, Mass.: Harvard University Press, 1968.

Kahn, Frank J., ed. *Documents of American Broadcasting*. New York: Appleton-Century-Crofts, 1972.

Katz, Elihu, and Tamas Szecsko, eds. *Mass Media and Social Change*. Beverly Hills, Calif.: Sage, 1981.

Klapper, Joseph T. *The Effects of Mass Media*. New York: Columbia University Press, 1949.

Kraus, Sidney, and Dennis Davis. *The Effects of Mass Communication on Political Behavior*. University Park: Pennsylvania State University Press, 1978.

Krieger, Susan. "Prospects for Communication Policy." *Policy Sciences* 2 (Summer 1971): 305–315.

Krippendorf, Klaus, ed. *Communications and Control in Society*. New York: Gordon and Breach, 1979.

Lang, Kurt, and Gladys Engel Lang. *Politics and Television*. Chicago: Quadrangle Books, 1968.

Lasswell, Harold D. *World Revolutionary Elites: Studies in Coercive Ideological Movements*. Cambridge, Mass.: MIT Press, 1965.

Lasswell, Harold D., and Abraham Kaplan. *Power and Society: A Framework for Political Inquiry*. New Haven, Conn.: Yale University Press, 1950.

Lasswell, Harold D., Daniel Lerner, and Hans Speier. *Propaganda and Communication in World History*. Vol. 1, *The Symbolic Instrument in Early Times*. Vol. 2, *Emergence of Public Opinion in the West*. Vol. 3, *A Pluralizing World in Formation*. Honolulu: University of Hawaii Press, for the East-West Center, 1979.

Lazarsfeld, Paul F., William H. Sewell, and Harold L. Wilensky, eds., *The Uses of Sociology*. New York: Basic Books, 1967.

Lazarsfeld, Paul F., and Frank N. Stanton, eds. *Communications Research, 1948–1949*. New York: Harper and Brothers, 1949.

Lefever, Ernest W. *TV and National Defense: An Analysis of CBS News, 1972–1973*. Boston, Va.: Institute for American Strategy, 1974.

Leites, Nathan. *A Study of Bolshevism*. Glencoe, Ill.: Free Press, for the Rand Corporation, 1953.

Lerner, Daniel. "Communication and Development." In *Communication Research: A Half-Century Appraisal*. Edited by Daniel Lerner et al. Honolulu: University of Hawaii Press, 1977.

———. *The Passing of Traditional Society: Modernizing the Middle East*. Glencoe, Ill.: Free Press, 1958.

———, ed. *Propaganda in War and Crisis*. New York: Stewart, 1951.

Lerner, Daniel, and Wilbur Schramm. *Communication and Change in Developing Countries*. Honolulu: East-West Center Press, 1967.

Lewin, L., ed. *Telecommunications: An Interdisciplinary Survey*. London: Artech House, 1979.

Lundberg, George A., Clarence C. Schrag, and Otto N. Larsen. "Mass Communication and Public Opinion." In *Sociology*. 4th ed. Edited by George A.

Lundberg, Clarence C. Schrag, and Otto N. Larsen. New York: Harper and Brothers, 1954.
McLuhan, Marshall. *Understanding Media: The Extensions of Man*. New York: New American Library, Signet, 1964.
McLuhan, Marshall, and Quentin Fiore. *War and Peace in the Global Village: An Inventory of Some of the Current Spastic Situations that Could be Eliminated by More Feed-forward*. New York: McGraw-Hill, 1968.
McQuail, Denis, ed. *Sociology of Mass Communications*. Baltimore, Md.: Penguin, 1972.
Merton, Robert K. *Mass Persuasion*. New York: Harper, 1946.
Merton, Robert K., and Paul F. Lazarsfeld, eds. *Continuities in Social Research: Studies in the Scope and Method of "The American Soldier."* Glencoe, Ill.: Free Press. 1950.
Mickiewicz, Ellen Propper. *Media and the Russian Public*. New York: Praeger, 1981.
Mills, C. Wright. *The Power Elite*. New York: Oxford University Press, 1959.
Mills, C. Wright, and I. L. Horowitz, eds. *Power, Politics, and People*. New York: Random House, Ballantine Books, 1963.
Moore, Wilbert E. "Sociology in Developing Areas." In *The Uses of Sociology*. Edited by Paul F. Lazarsfeld, William H. Sewell, and Harold L. Wilensky. New York: Basic Books, 1967.
Nafziger, Ralph O., ed. *Introduction to Mass Communications Research*. Baton Rouge: Louisiana State University Press, 1958.
Oettinger, Anthony G., Paul J. Berman, and William H. Read. *High and Low Politics: Information Resources for the 80's*. Cambridge, Mass.: Ballinger, 1977.
Parsons, Talcott. *Essays in Sociological Theory*. Glencoe, Ill.: Free Press, 1949.
———. *The Social System*. Glencoe, Ill.: Free Press, 1951.
———. *The Sociological Theory and Modern Society*. New York: Free Press, 1967.
Pastore, J. O. *The Story of Communications*. New York: Macfadden-Bartell, 1964.
Paulu, Burton. *Radio and Television Broadcasting in Eastern Europe*. Minneapolis: University of Minnesota Press, 1974.
Pool, Ithiel de Sola. "Technology and Policy in the Information Age." In *Communication Research: A Half-Century Appraisal*. Edited by Daniel Lerner et al. Honolulu: University of Hawaii Press, 1977.
———, ed. *Talking Back: Citizen Feedback and Cable Technology*. Cambridge, Mass.: MIT Press, 1973.
Pool, Ithiel de Sola, and Wilbur Schramm, eds. *Handbook of Communications*. Chicago: Rand McNally, 1973.
Pool, Ithiel de Sola, Philip J. Stone, and Alexander Szalai. *Communications, Computers, and Automation for Development*. New York: Unitar, 1971.
Pye, Lucian. *Aspects of Political Development*. Boston: Little, Brown, 1966.
———, ed. *Communications and Political Development*. Princeton, N.J.: Princeton University Press, 1963.
Pye, Lucian, and Sidney Verba, eds. *Political Culture and Political Development*. Princeton, N.J.: Princeton University Press, 1965.
Read, William H. *America's Mass Media Merchants*. Baltimore: Johns Hopkins University Press, 1976.
Rogers, Everett M., ed. *Communication and Development: Critical Perspectives*. Sage Contemporary Social Science Issues, no. 32. Beverly Hills, Calif.: Sage, 1976.

Rogow, Arnold A., ed. *Politics, Personality, and Social Science in the Twentieth Century: Essays in Honor of Harold D. Lasswell*. Chicago: University of Chicago Press, 1969.

Schramm, Wilbur. *Mass Media and National Development: The Role of Information in the Developing Countries*. Stanford, Calif.: Stanford University Press, 1964, 1967.

———. *Men, Messages, Media: A Look at Communications*. New York: Harper & Row, 1973.

———, ed. *Mass Communications*. Urbana: University of Illinois Press, 1948, 1960.

Skornia, Harry J. *TV and Society*. New York: McGraw-Hill, 1965.

Smith, Bruce Lannes, and Chitra M. Smith, eds. *International Communication and Political Opinion: A Guide to the Literature*. Princeton, N.J.: Princeton University Press, 1956. (Prepared for the Rand Corporation by the Bureau of Social Science Research)

Stanton, Frank. "Parallel Paths." In *Culture for the Millions: Mass Media in Modern Society*. Edited by Norman Jacobs. Boston: Beacon Press, 1959.

Teheranian, M., F. Hakimzadeh, and M. L. Vidale. *Communication Policy for National Development*. London: Routledge & Kegan Paul, 1977.

Tunstall, Jeremy. *The Media Are American: Anglo-American Media in the World*. New York: Columbia University Press, 1977.

———, ed. *Media Sociology: A Reader*. Urbana: University of Illinois Press, 1970.

Whitton, John B., and Arthur Larson. *Propaganda: Towards Disarmament in the War of Words*. Dobbs Ferry, N.Y.: Oceana Publications, 1964.

Winick, Charles. *Taste and the Censor on Television*. New York: Fund for the Republic, 1965.

International Law and International Broadcasting

Studies Published Before 1970

Adams, Thomas R. "The Outer Space Treaty: An Interpretation in Light of the No-Sovereignty Provision." *Harvard International Law Journal* 9 (Winter 1968): 140–157.

Bloomfield, Lincoln P. "The Prospects for Law and Order." In *Outer Space: Prospects for Man and Society*. Edited by Lincoln P. Bloomfield. Englewood Cliffs, N.J.: Prentice-Hall, for the American Assembly, Columbia University, 1962.

Christol, Carl Q. *The International Law of Outer Space*. U.S. Naval War College International Law Studies. Washington, D.C.: Government Printing Office, 1966.

Cooper, John Cobb. *Explorations in Aerospace Law: Selected Essays by John Cobb Cooper*. Edited by Ivan A. Vlasic. Montreal: McGill University Press, 1968.

Crane, R. D. "The Beginnings of Marxist Space Jurisprudence." *American Journal of International Law* 56 (July 1963): 616–625.

Doyle, Stephen E. "An Analysis of the Socialist States' Proposal for Intersputnik: An International Communication Satellite System." *Villanova Law Review* 15 (Fall 1969): 83–105.

———. "Communication Satellites: International Organizations for Development and Control." *California Law Review*. 55 (May 1967): 431–448.

Fawcett, J. E. S. *International Law and the Uses of Outer Space*. Manchester: Manchester University Press, 1968.

Gal, Gyula. *Space Law*. Leyden: Sijthoff; Dobbs Ferry, N.Y.: Oceana Publications, 1969.

Gotlieb, A. E., and C. M. Dalfen. "Direct Satellite Broadcasting: A Case Study in the Development of the Law of Space Communications." In *Canadian Yearbook of International Law*. Vol. 7. Vancouver: University of British Columbia Press, 1969.

Havighurst, C. C., ed. *International Control of Propaganda*. Dobbs Ferry, N.Y.: Oceana Publications, for the Duke University Law School, 1967.

Hoffman, Stanley, and Karl W. Deutsch, eds. *The Relevance of International Law: Essays in Honor of Leo Gross*. Cambridge, Mass.: Schenkman, 1968.

Jenks, C. Wilfred. *Space Law*. New York: Praeger, 1965.

Jessup, Philip C. *A Modern Law of Nations: An Introduction*. New York: Macmillan, 1948, 1968.

Jessup, Philip C., and Howard J. Taubenfeld. *Controls for Outer Space and the Antarctic Analogy*. New York: Columbia University Press, 1959.

Johnson, Roger Maurice. "Soviet Doctrines of International Law: General International Law and Socialist International Law." Geneva: Graduate Institute of International Studies, 1974, Mimeo.

Josephson, Matthew. *Empire of the Air: Juan Trippe and the Struggle for World Airways*. New York: Harcourt, Brace, 1944.

Kelson, Hans. "The Essence of International Law." In *The Relevance of International Law: Essays in Honor of Leo Gross*. Edited by Stanley Hoffman and Karl W. Deutsch. Cambridge, Mass.: Schenkman, 1968.

Laskin, Paul L. "Background paper." In *Communicating by Satellite*. Report of the Twentieth Century Fund Task Force on International Satellite Communcations. New York: Twentieth Century Fund, 1969.

Lissitzyn, Oliver James. *International Law Today and Tomorrow*. Dobbs Ferry, N.Y.: Oceana Publications, 1965.

McDougal, M. S., Harold D. Lasswell, and I. A. Vlasic. *Law and Public Order in Space*. New Haven, Conn.: Yale University Press, 1963.

McWhinney, Edward. *"Peaceful Coexistence" and Soviet-Western International Law*. Leyden: Sijthoff, 1964.

Moronoff, Jerome. *World Peace Through Space Law*. Charlottesville, Va.: Michie, 1967.

Schwartz, Herman. "Governmentally Appointed Directors in a Private Corporation: The Communications Satellite Act of 1962." *Harvard Law Review* 79 (December 1965): 350–364.

Smith, Delbert. *International Telecommunication Control*. Leyden: Sijthoff, 1969.

Vlasic, Ivan A. "The Space Treaty: A Preliminary Evaluation." *California Law Review* 55 (May 1967): 507–540.

Studies Published Since 1970

Alexandrowicz, Charles Henry. *The Law of Global Communications*. New York: Columbia University Press, 1971.

Andhyarujna, Firoze. "Direct Satellite Telecommunication and Space Law." Paper presented at the Nineteenth Colloquium on the Law of Outer Space, International Institute of Space Law, 1977.

Bhatt, Saligram. *Legal Controls of Outer Space: Law, Freedom, and Responsibility*. New Delhi: S. Chand, 1973.

Bocksteigel, Karl-Heinz. "Arbitration and Adjudication Regarding Activities in Outer Space." *Journal of Space Law* 6 (Spring 1978): 3–18.

Brown, Elton Rex III. "DBS and Freedom of Speech." *California Western International Law Journal* 4 (Spring 1974): 374–392.

Burling, James C. et al. "Direct Satellite Broadcasting and the First Amendment." *Harvard International Law Journal* 15 (Summer 1974): 514–527.

Busak, Jan. "Perspectives de la télévision et de la radiodiffusion directe par satellite." In *Space Law Perspectives: Commentaries Based on Volumes 1–15 (1957–1972) of the Colloquia on the Law of Outer Space*. Edited by Mortimer D. Schwartz. South Hackensack, N.J.: Rothman, 1976.

Chapman, John H., and Gabriel I. Warren. "Direct Broadcast Satellites: The ITU, UN, and the Real World." In *Annals of Air and Space Law*. Vol. 4. Edited by N. M. Matte. Toronto: Carswell, 1979.

Chayes, Abram. "Unilateralism in the United States Satellite Communications Policy." In *The International Law of Communications*. Edited by Edward McWhinney. Dobbs Ferry, N.Y.: Oceana Publications, 1971.

Chayes, Abram, and Leonard Chazen. "Policy Problems in Direct Broadcasting from Satellites." *Stanford Journal of International Studies* 5 (June 1970): 4–20.

Chayes, Abram, and Paul Laskin. *Direct Broadcasting from Satellites: Policies and Problems*. A Report of the Panel on International Telecommunications Policy. Studies in Transnational Legal Policy, no. 7, Washington, D.C.: American Society of International Law, 1975.

Chayes, Abram et al. *Satellite Broadcasting*. London: Oxford University Press, 1973.

Christol, Carl Q. "The International Telecommunication Union and the International Law of Outer Space." Paper presented at the Twenty-first Colloquium on the Law of Outer Space, International Institute of Space Law, 1979.

———. *Modern International Law of Outer Space*. New York: Pergamon Press, 1982.

———. "The 1974 Brussels Convention Relating to the Distribution of Program-Carrying Signals Transmitted by Satellite: An Aspect of Human Rights." *Journal of Space Law* 6 (Spring 1978): 19–35.

Cocca, Aldo A. "Extract from the Introductory Report of Part II (Legal Problems of Telecommunications, including Direct Broadcasting from Outer Space." In *Space Law Perspectives: Commentaries Based on Volumes 1–15 (1957–1972) of the Colloquia on the Law of Outer Space*. Edited by Mortimer D. Schwartz. South Hackensack, N.H.: Rothman, 1976.

Colino, Richard R. "Challenges Facing Intelsat." *Transnational Data and Communications Report* 9 (June 1986): 13–15.

———. "Reply to Mr. Vereshchetin." In *Space Law Perspectives: Commentaries Based on Volumes 1–15, (1957–1972) of the Colloquia on the Law of Outer Space*. Edited by Mortimer D. Schwartz. South Hackensack, N.J.: Rothman, 1976.

Dalfen, C. M. "Direct Satellite Broadcasting: Towards International Arrangements to Transcend and Marshal the Political Realities." *University of Toronto Law Journal* 20 (1970): 366–374.

———. "The International Legislative Process: Direct Broadcasting and Remote Earth Sensing by Satellite Compared." In *Canadian Yearbook of International Law*. Vancouver: University of British Columbia Press, 1972.

Dauses, Manfred. "Direct Television Broadcasting by Satellites and Freedom of Information." *Journal of Space Law* 3 (Spring and Fall 1975): 59–71.

Diamond, Edwin, Norman Sandler, and Milton Mueller. *Telecommunication in Crisis: The First Amendment, Technology, and Deregulation*. Washington, D.C.: Cato Institute, 1983.

Dickson, B. H. "Effects of 1977 ITU World Administrative Radio Conference on the Formulation of UN Draft Principles on Direct Broadcast Satellites." In *Annals of Air and Space Law*. Vol. 2. Toronto: Carswell, 1977.

Dudakov, B. G. "Some International Legal Issues on the Direct Television Broadcast Satellites." In *Space Law Perspectives: Commentaries Based on Volumes 1–15 (1957–1972) of the Colloquia on the Law of Outer Space*. Edited by Mortimer D. Schwartz. South Hackensack, N.J.: Rothman, 1976.

Dugmore, K. C. "Newest Frontier in Communications: The Direct Broadcast Satellite." *Air Force Law Review* 13 (Fall 1971): 259–269.

Eek, Hilding A. "Principles Governing the Use of the Mass Media as Defined by the United Nations and UNESCO." In *National Sovereignty and International Communication*. Edited by Kaarle Nordenstreng and Herbert I. Schiller. Norwood, N.J.: Ablex, 1979.

Emerson, Thomas I. *The System of Freedom of Expression*. New York: Random House, 1970.

Emin, V. "International Law: Direct Broadcasting Satellites." *International Affairs* [Moscow], no. 12 (December 1971): 116–117.

Errera, Roger. "Problems Raised by the Content of Television Programs Transmitted by Telecommunication Satellites." In *The International Law of Communications*. Edited by Edward McWhinney. Dobbs Ferry, N.Y.: Oceana Publications, 1971.

Finch, Edward R., Jr., and Amanda L. Moore. "Outer Space Law and the Global Community." *The International Lawyer* 8 (October 1974): 752–771.

Freeman, Jennifer. "Toward the Free Flow of Information: Direct Television Broadcasting via Satellite." *Journal of International Law and Economics* 13, no. 2 (1979): 329–366.

Friedman Memorial Program. "The Control of Program Content in International Communications: A Discussion of General Principles" (moderator: Fred Friendly; discussants: Abram Chayes, Frank Shakespeare, Harry Olsson, and David Webster). *Columbia Journal of Transnational Law* 13 (1974): 40–67.

———. "The Control of Program Content in International Communications: Direct Satellite Broadcasting and the Third World" (discussants: Luiz Felipe de Seixas Corrêa, Nandasiri Jasentuliyana, Abram Chayes, Josef C. Nichols, and Irwin Singer). *Columbia Journal of Transnational Law* 13 (1974): 68–81.

Frutkin, Arnold W. "Direct/Community Broadcast Using Space Satellites." *Journal of Space Law* 3 (Spring and Fall 1975): 17–24.

Gal, Gyula. "Positive and Negative Aspects of the Law of Direct Television Broadcasting from Space." In *Proceedings of the Sixteenth Colloquium on the Law of Outer Space*. International Institute of Space Law, 1973.

Galloway, Eilene. "Consensus Decisionmaking by the Unted Nations Committee on the Peaceful Uses of Outer Space." *Journal of Space Law* 7 (Spring 1979): 3–13.

———. "Direct Broadcast Satellites and Space Law." *Journal of Space Law* 3 (Spring and Fall, 1975): 3–15.

Galloway, Jonathan. "The Current Status of the Controversy over the Geostationary Orbit." Paper presented at the Twentieth Colloquium on the Law of Outer Space, International Institute of Space Law, 1978.

Gehrig, James J. "Broadcasting Satellites—Prospects and Problems." *Journal of Space Law* 3 (Spring and Fall 1975): 25–38.

Geston, M. S. "Direct Broadcasting from Satellite: The Case for Regulation." *New York University Journal of International Law and Politics* (Spring 1970): 73–93.

Goedhuis, D. "Influence of the Conquest of Outer Space on National Sovereignty: Some Observations." *Journal of Space Law* 6 (Spring 1978): 37–46.

Gold, Martin E. "Direct Broadcast Satellites: Implications for Less Developed Countries and for World Order." *Virginia Journal of International Law* 12 (December 1971): 66–90.

Gorbiel, Andrzej. "The Legal Status of the Geostationary Orbit: Some Remarks." *Journal of Space Law* 6 (Fall 1978): 171–177.

Gotlieb, Allan, Charles Dalfen, and Kenneth Katz. "The Transborder Transfer of Information by Communications and Computer Systems: Issues and Approaches to Guiding Principles." *American Journal of International Law* 68 (April 1974): 227–257.

Gross, Leo. "Some International Law Aspects of the Freedom of Information and the Right to Communicate." In *National Sovereignty and International Communication*. Edited by Kaarle Nordenstreng and Herbert I. Schiller. Norwood, N.J.: Ablex, 1979.

Haanappel, P. C. "The Stagnating Development of International Space Law and its Causes." Paper presented at the Twenty-first Colloquium on the Law of Outer Space, International Institute of Space Law, 1979.

Hosenball, S. Neil. "The United Nations Committee on the Peaceful Uses of Outer Space: Past Accomplishments and Future Challenges." *Journal of Space Law* 7 (Fall 1979): 95–113.

Ickowitz, Allan H. "The Role of the ITU in the Settlement of Harmful Interference Disputes." *Columbia Journal of Transnational Law* 13, no. 1 (1974): 82–97.

Jasentuliyana, Nandasiri, and Roy S. K. Lee, eds. *Manual on Space Law*. 2 vols. Dobbs Ferry, N.Y.: Oceana Publications, 1979.

Jones, Erin Bain. *Earth Satellite Telecommunications Systems and International Law*. Austin: University of Texas Press, 1970.

Jourakov, V. G. "Legal Regime of Social Consequences of Direct Television Broadcasting Through Artificial Earth Satellites (DTB)." Paper presented at the Sixteenth Colloquium on the Law of Outer Space, International Institute of Space Law, 1973.

Journal of Space Law. Published since 1973 by University of Mississippi Law School.

Kamenetskaya, E. "Subjects of International Space Law." Paper presented at the Twenty-first Colloquium on the Law of Outer Space, International Institute of Space Law, 1979.

Kolossov, Yuri. "Legal Consequences of 'Spill-over,' Resulting from Satellite Direct Broadcasting." In *Space Law Perspectives: Commentaries Based on Volumes 1–15 (1957–1972) of the Colloquia on the Law of Outer Space*. Edited by Mortimer D. Schwartz. South Hackensack, N.J.: Rothman, 1976.

———. "Space and International Law." *International Affairs* [Moscow], no. 8 (August 1977): 54–61.

Laskin, Paul L., and Abram Chayes. "International Satellite Controversy." *Society* 12 (September–October 1975): 30–40.

Lay, S. Houston, and Howard J. Taubenfeld. *The Law Relating to Activities of Man in Space*. American Bar Foundation Study. Chicago: University of Chicago Press, 1970.

Lay, S. Houston et al. "Preliminary Draft of a Study of Censorship Provisions of a Proposed Telecommunications Satellite Treaty and the Constitution of the USA." In *Proceedings of the Seventeenth Colloquium on the Law of Outer Space*. International Institute of Space Law, 1974.

Leive, David M. *International Telecommunications and International Law: The Regulation of the Radio Spectrum*. Dobbs Ferry, N.Y.: Oceana Publications, 1970.

McDougal, Myres S., Harold D. Lasswell, and Lung-chu Chen. *Human Rights and World Public Order: The Basic Policies of an International Law of Human Dignity*. New Haven, Conn.: Yale University Press, 1980.

McWhinney, Edward. "The Antinomy of Policy and Function in the Institutionalization of International Telecommunications Broadcasting." *Columbia Journal of Transnational Law* 13 (1974): 3–39.

———, ed. *The International Law of Communications*. Dobbs Ferry, N.Y.: Oceana Publications, 1971.

Magno, Pompeo, "The Rules Governing the 'Direct Broadcasting Satellite' at the U.N." In *Space Law Perspectives: Commentaries Based on Volumes 1–15 (1957–1972) of the Colloquia on the Law of Outer Space*. Edited by Mortimer D. Schwartz. South Hackensack, N.J.: Rothman, 1976.

Matte, N. M. *Aerospace Law: From Scientific Exploration to Commercial Utilization*. Toronto: Carswell, 1979.

———, ed. *Annals of Air and Space Law*. Vol 4. Toronto: Carswell, 1979.

Mizrack, Richard. "The Intelsat Definitive Arrangements." *Journal of Space Law* 1 (Fall 1973): 129–138.

Nordenstreng, Kaarle, and Herbert I. Schiller, eds. *National Sovereignty and International Communication*. Norwood, N.J.: Ablex, 1979.

Nozari, Fariborz. *The Law of Outer Space*. Stockholm: Norstedt, 1973.

O'Brien, William V. "International Propaganda and Minimum World Public Order." In *Intercommunications Among Peoples*. Edited by Michael E. Prosser. New York: Harper & Row, 1973.

Ogunbanwo, Ogunsola O. *International Law and Outer Space Activities*. The Hague: Martinus Nijhoff, 1975.

Patermann, Christian. "Applicable Law in Cases of Tort Damages Caused by Direct Broadcast Satellites." *Journal of Space Law* 3 (Spring and Fall 1975): 47–54.

Pikus, Irwin M. "Legal Implications of DBS Technology." *Journal of Space Law* 3 (Spring and Fall 1975): 39–46.

Piradov, A. S., ed. *International Space Law*. Moscow: Progress Publishers, 1976.

Plasse, J. M. "Recent Developments in the Law of Direct Broadcast Satellites." *Brooklyn Journal of International Law* 2 (Fall 1975): 139–163.

Ploman, Edward, "Some Observations on Legal Problems in Connection with Satellite Communications." In *Space Law Perspectives: Commentaries Based on Volumes 1–15 (1957–1972) of the Colloquia on the Law of Outer Space*. Edited by Mortimer D. Schwartz. South Hackensack, N.J.: Rothman, 1976.

Powell, J. T. "DBS: The Conceptual Conveyance of the Free Flow of Information and National Sovereignty." *California Western International Law Journal* 6 (Winter 1975): 1–40.

Price, Monroe E. "First Amendment and Television Broadcasting by Satellite." *UCLA Law Review* 23 (June 1976): 879–903.

Quinney, Richard, and Piers Beirne. *Marxism and Law*. New York: Wiley, 1982.

Rankin, Clyde E. III. "Utilization of the Geostationary Orbit—A Need for Orbital Allocation?" *Columbia Journal of Transnational Law* 13, no. 1 (1974): 98–113.

Ruddy, F. S. "The Freedom of Information and Direct Broadcast Satellites." Paper presented at the Sixteenth Colloquium on the Law of Outer Space, International Institute of Space Law, 1973.

Sarkar, Subrata, "Introduction to Space Telecommunications Relations." Paper presented at the Twentieth Colloquium on the Law of Outer Space, International Institute of Space Law, 1978.

———. "Requirements for Establishing a Broadcast-satellite Service." Paper presented at the Sixteenth Colloquium on the Law of Outer Space, International Institute of Space Law, 1973.

Sarnoff, R. W. "One World of Communications." *International Lawyer* 6 (January 1972): 42–47.

Schwartz, Mortimer D., ed. *Space Law Perspectives: Commentaries Based on Volumes 1–15 (1957–1972) of the Colloquia on the Law of Outer Space*. South Hackensack, N.J.: Rothman, 1976. Sponsored by the International Institute of Space Law, of the International Astronautical Federation.

Schwarzenberger, Georg. *The Dynamics of International Law*. Milton, Eng.: Professional Books, 1976.

Shillinglaw, Thomas L. "The Soviet Union and International Satellite Telecommunications." *Stanford Journal of International Studies* 5 (June 1970): 199–226.

Smith, Delbert D. "Educational Satellite Telecommunication: The Challenge of a New Technology." *Bulletin of the Atomic Scientists* 27 (April 1971): 14–18.

Socol, Scott K. "COMSAT's First Decade: Difficulties in Interpreting the Communications Satellite Act of 1962." *Georgia Journal of International and Comparative Law* 7 (1977): 678–692.

Tunkin, G. I. *Theory of International Law*. Cambridge, Mass.: Harvard University Press, 1974.

University of Toronto Law Journal, in association with Canadian Department of Communications and the International Law Association. "Proceedings of Conference on the Legal Problems of International Communications: The Legal Problems of International Telecommunications with Special Reference to Intelsat" (discussants: Alan Beesley, Edward McWhinney, Dallas W. Smythe, Barry Mawhinney, and A. E. Gotlieb). *University of Toronto Law Journal* 20 (1970): 287–313.

———. "Proceedings of Conference on the Legal Problems of International Communications: International Legal Problems of Direct Satellite Broadcasting" (discussants: C. M. Dalfen, André Bissonnette, Pierre Juneau, and Ivan Vlasic). *University of Toronto Law Journal* 20 (1970): 314–332.

Vereshchetin, V. S., "Legal Questions of Satellites Telecommunications at the Washington Intelsat Conference." In *Space Law Perspectives: Commentaries Based on Volumes 1–15 (1957–1972) of the Colloquia on the Law of Outer Space*. Edited by Mortimer D. Schwartz. South Hackensack, N.J.: Rothman, 1976.

Wiewiorowska, Krystyna. "Outer Space and International Law." *Polish Perspectives* 23 (July–August 1980): 11–14.

Zhukov, G. P., "International Law Problems of Direct Television Broadcasts from Satellites." Paper presented at the Eighteenth Colloquium on the Law of Outer Space, International Institute of Space Law, 1976.

Critical and Marxist Approaches to Mass Communications

Adorno, Theodor. "Television and the Patterns of Mass Culture." In *Mass Communications*. Edited by Wilbur Schramm. Urbana: University of Illinois Press, 1948, 1960.

Afanasyev, V. *Social Information and the Regulation of Social Development*. Moscow: Progress Publishers, 1978.

Arbatov, Georgi. *The War of Ideas in Contemporary International Relations*. Moscow: Progress Publishers, 1973.

Arditi, Rita, Pat Brennan, and Steve Cavrak, eds. *Science and Liberation*. Boston. South End Press, 1980.

Aronowitz, Stanley. *False Promises: The Formation of American Working Class Consciousness*. New York: McGraw-Hill, 1973.

Aronson, James. *Packaging the News: A Critical Survey of Press, Radio, and TV*. New York: International Publishers, 1971.

Baluyev, Boris. *Lenin and the Bourgeois Press*. Moscow: Progress Publishers, 1983.

Beltran, Luis Ramiro, and Elizabeth Fox de Cardona. "Latin America and the United States: Flaws in the Free Flow of Information." In *National Sovereignty and International Communication*. Edited by Kaarle Nordenstreng and Herbert I. Schiller. Norwood, N.J.: Ablex, 1979.

Benjamin, Walter. "A Radio Talk on Brecht." *New Left Review*, no. 123 (September–October 1980): 92–96.

———. *Reflections*. New York: Harcourt Brace Jovanovich, 1979.

———. "The Storyteller and Artisan Cultures." In *Critical Sociology*. Edited by Paul Connerton. Baltimore: Penguin, 1976.

Bernal, J. D. *Science in History*. New York: Hawthorn Books, 1965.

———. *Science and Industry in the 19th Century*. London: Routledge & Kegan Paul, 1953.

———. *The Social Function of Science*. New York: Macmillan, 1939.

———. *World Without War*. London: Routledge & Kegan Paul, 1958.

Biryukov, N. S. *Television in the West and Its Doctrines*. Moscow: Progress Publishers, 1981.

Blackburn, Robin, ed. *Ideology in Social Science: Readings in Critical Social Theory*. New York: Random House, Vintage Books, 1972.

Bobrakov, Yuri. "War Propaganda: A Serious Crime Against Humanity." In *Intercommunications Among Peoples*. Edited by Michael H. Prosser. New York: Harper & Row, 1973.

Brecht, Bertolt. *Sur le cinema*. Paris: l'Arche. 1970.

Brenkman, John. "Mass Media: From Collective Experience to the Culture of Privatization." *Social Text: Theory/Culture/Ideology*, no. 1. (Winter 1979): 94–109.

Brown, Michael E. "Sociology as Critical Theory." 1977. Mimeo.

Bugrov, Ye. *The U.S. Military–Industrial Complex Is a Threat to Peace*. International Peace and Disarmament Series. Moscow: Nauk, 1984.

Bunce, Richard. *Television in the Corporate Interest*. New York: Praeger, 1976.

Burton, Julianne, and Jean Franco. "Culture and Imperialism." *Latin American Perspectives* 5 (Winter 1978): 2–12.

Carmo, Alberto. "International Telephone and Telegraph Corporation—A Gigantic Multinational Octopus." *Democratic Journalist* [Prague], no. 9 (1973): 15–18.

Cazeneuve, Jean. *Les pouvoirs de la télévision*. Paris: Gallimard, 1970.

Chasin, Barbara, and Gerald Chasin. *Power and Ideology: A Marxist Approach to Political Sociology*. Cambridge, Mass.: Schenkman, 1974.

Cherminina, A. P. "The Concept of Freedom in Marxist-Leninist Ethics." In *Intercommunications Among Peoples*. Edited by Michael H. Prosser. New York: Harper & Row, 1973.

Culture in the Age of Mass Media. Special issue. *Latin American Perspectives* 5 (Winter 1978).

Cunningham, Ann Marie et al. *Techno/Peasant Survival Manual*. New York: Bantam Books, 1980.

de la Haye, Yves, ed. *Marx and Engels on the Means of Communication—The Movement of Commodities, People, Information, and Capital: A Selection of Texts*. New York: International General, in cooperation with International Mass Media Research Center (Paris), 1980.

Department of Journalism, Moscow University. "The Capitalist Mass Communication Media." *World Marxist Review* 16 (August 1973): 131–137.

Dorfman, Ariel. *The Empire's Old Clothes: What the Lone Ranger, Babar, and Other Innocent Heroes Do to Our Minds*. New York: Pantheon, 1983.

Dorfman, Ariel, and Armand Mattelart. *How to Read Donald Duck: Imperialist Ideology in the Disney Comic*. New York: International General, 1975.

Dreyer, O. *Cultural Changes in Developing Countries*. Problems of the Third World series. Moscow: Progress Publishers, 1976.

Eagleton, Terry. *Marxism and Literary Criticism*. London: Methuen, 1976.

Ehrlich, Howard J. "The Politics of News Media Control." *Insurgent Sociologist* 4 (Summer 1974).

Enzensberger, Hans Magnus. *The Consciousness Industry: On Literature, Politics, and the Media*. New York: Seabury Press, 1974.

———. "Constituents of a Theory of the Media." *New Left Review*, no. 64 (November–December 1970): 13–36.

Epstein, S. "Imperialism and Manipulating with Public Opinion." *Democratic Journalist* [Prague], no. 9 (1973): 4–5.

Ewen, Stuart. *Captains of Consciousness: Advertising and the Social Roots of the Consumer Culture*. New York: McGraw-Hill, 1976.

Ewen, Stuart, and Elizabeth Ewen. *Channels of Desire: Mass Images and the Shaping of American Consciousness*. New York: McGraw-Hill, 1982.

Finkelstein, Sidney. *Sense and Nonsense of McLuhan*. New York: International Publishers, 1969.

Foucault, Michel. *The Archeology of Knowledge and the Discourse on Language*. New York: Harper & Row, 1972.

Galtung, Johan. "Data Collection." In *Sociological Perspectives*. Edited by Jeremy Tunstall et al. Baltimore: Penguin, 1971.

Garnham, Nicholas. "Contribution to a Political Economy of Mass Communication." *Media, Culture and Society* 1 (1979): 123–146.

———. *Structures of Television*. London: British Film Institute Educational Advisory Service, 1973.

Gitlin, Todd. *Inside Prime Time*. New York: Pantheon Books, 1983.

———. "Media Sociology: The Dominant Paradigm." *Theory and Society* 6 (September 1978): 205–253.

———. *The Whole World Is Watching: Mass Media in the Making and Unmaking of the New Left*. Berkeley: University of California Press, 1980.

Gramsci, Antonio. *The Modern Prince and Other Writings*. New York: International Publishers, 1970.

———. *Selections from the Prison Notebooks*. Edited by Quinton Hoare and Geffrey Nowell Smith. New York: International Publishers, 1971.

Gutto, S. B. O. "Captivity of the Press." *Monthly Review* 34 (January 1983): 31–41.

Habermas, Jurgen. *Communication and the Evolution of Society*. Boston: Beacon Press, 1979.

Hamelink, Cees J. "The New Information Order and the Industrial State." In *Peace and the Sciences*. Report of International Institute for Peace, Vienna, on scientific symposium held in Orivesi, Tampere, Finland, November 9–11, 1977.

International Institute for Peace, Vienna. "The Role of the Mass Media in Maintaining and Strengthening Peace." In *Peace and the Sciences*. Report of scientific symposium held in Orivesi, Tampere, Finland, November 9–11, 1977.

IMMRC, ed. *Marxism and the Mass Media: Towards a Basic Bibliography*. 4 vols. Bagnolet, France: International Mass Media Research Center, 1976, 1978, 1980, 1983.

Jameson, Fredric. "Postmodernism, or the Cultural Logic of Late Capitalism." *New Left Review*, no. 146 (July–August 1984): 53–92.

Kashlev, Y. "The 'Information Explosion' and Information Imperialism." *International Affairs* (December 1984): 58–65.

Man in the "Industrial Society." Special issue. *World Marxist Review* 8 (March 1965).

Mannheim, Karl. *Man and Society in an Age of Reconstruction*. New York: Harcourt, Brace, 1940.

Marx, Karl, and Friedrich Engels. *On Literature and Art*. Moscow: Progress Publishers, 1978.

Mattelart, Armand. "Mass Communications in the Third World: Theory and Practice." *Latin American Perspectives* 5 (Winter 1978): 13–34.

———. *Mass Media, Ideologies, and the Revolutionary Movement*. Atlantic Highlands, N.J.: Humanities Press, 1980.

———. *Transnationals and the Third World: The Struggle for Culture*. South Hadley, Mass.: Bergin and Garvey, 1983.

Mattelart, Armand, Xavier Delcourt, and Michelle Mattelart. *International Image Markets: In Search of an Alternative Perspective*. London: Comedia, 1984.

Mattelart, Armand, and Seth Siegelaub, eds. *Communication and Class Struggle*. 2 vols. New York: International General, 1979, 1983.

Mosco, Vincent, and Andrew Herman. "Radical Social Theory and the Communications Revolution." Paper presented at the Fourth Annual Conference on the Current State of Marxist Theory, University of Louisville, 1979.

Mosco, Vincent, and Janet Wasko, eds. *The Critical Communications Review*. Vol. 2. Norwood, N.J.: Ablex, 1984.

Mueller, Claus. *The Politics of Communication: A Study in the Political Sociology of Language, Socialization, and Legitimation*. New York: Oxford University Press, 1973.

Nikolayev, Lev. *Whose Interests Does Psychological Warfare Serve?* Moscow: Novosti, 1983.

Nordenstreng, Kaarle. "Bitter Lessons." *Journal of Communications* 34 (Winter 1984): 138–142.

———. "From Mass Media to Mass Consciousness." In *Mass Media Policies in Changing Cultures*. Edited by George Gerbner. New York: Wiley, 1977.

Panfilov, A. *Broadcasting Pirates or Abuse of the Microphone: An Outline of External Political Radio Propaganda by the USA, Britain and the FRG*. Moscow: Progress Publishers, 1981.

Parenti, Michael. "How Monopoly Dominates Culture." *Political Affairs: Journal of Marxist Thought* 64 (March 1985): 3–12.

———. *Inventing Reality: The Politics of Mass Communication*. New York: St. Martin's, 1986.

Pasquali, Antonio. "Understanding Communication or the Media?" *Cultures* [UNESCO] 6 no. 3. (1979): 11–28.

Pietilä, Veikko. "On the Scientific Status and Position of Communication Research." Institute of Journalism and Mass Communication, Tampere, Finland, 1977. Mimeo.

Pisarek, Walery. "Marxist-Leninist Theory and Methodology of Media Research: Principles and Directives." In *Marxist-Leninist Theory and Methodology of Media Research*. Edited by Walery Pisarek and Tomasz Goban-Klas. Cracow: Press Research Center, 1978.

Radical Science Collective, ed. *Making Waves: The Politics of Communications*. London: Free Association Books, 1985.

Rapping, Elayne. "Processed Ideas and Packaged Dreams: The Manufacturing and Marketing of American Reality." New American Movement, n.d. Mimeo.

Rude, George. *Ideology and Popular Protest*. New York: Pantheon Books, 1980.

Salinas, Raquel and Leena Paldán. "Culture in the Process of Dependent Development: Theoretical Perspectives." In *National Sovereignty and International Communication*. Edited by Kaarle Nordenstreng and Herbert I. Schiller. Norwood, N.J.: Ablex, 1979.

Sallach, David L. "Class Domination and Ideological Hegemony." In *The TV Establishment: Programming for Power and Profit*. Edited by Gaye Tuchman. Englewood Cliffs, N.J.: Prentice-Hall, 1974.

Schiller, Herbert I. *Communication and Cultural Domination*. White Plains, N.Y.: International Arts and Sciences Press, 1976.

———. "Decolonization of Information: Efforts Toward a New Information Order." *Latin American Perspectives* 5 (Winter 1978): 35–48.

———. "Freedom from the 'Free Flow.'" *Journal of Communication* 24 (1974): 110–117.

———. *Information and the Crisis Economy*. Norwood, N.J.: Ablex, 1984.

———. *Mass Communication and American Empire*. New York: Kelley, 1969.

———. *The Mind Managers*. Boston: Beacon Press, 1973.

———. *Who Knows: Information in the Age of the Fortune 500*. Norwood, N.J.: Ablex, 1981.

Selsam, Howard et al., eds. *Dynamics of Social Change*. New York: International Publishers, 1970.

Sherman, Howard J. "'Technology vis-à-vis Institutions': A Marxist Commentary." *Journal of Economic Issues* 13 (March 1979): 175–191.

Smith, Anthony. *The Geopolitics of Information: How Western Culture Dominates the World*. New York: Oxford University Press, 1980.

Smith, Henry. "Goebbels in Space: Government Use of Telecommunications." *Cineaste* 5 (1973): 37–44.

Smythe, Dallas W. *Dependency Road: Communications, Capitalism, Consciousness, and Canada*. Norwood, N.J.: Ablex, 1981.

Snedeker, George. "Western Marxism and the Problem of History." *Psychology and Social Theory*, no. 4 (1984): 41–50.

Swingewood, Alan. *The Myth of Mass Culture*. Atlantic Highlands, N.J.: Humanities Press, 1977.

Szecsko, Tamas. "The Development of a Socialist Communication Theory." In *Mass Media Policies in Changing Cultures*. Edited by George Gerbner. New York: Wiley, 1977.

Terrar, Toby. "New Information Order and U.S. Free Press Tradition." *Democratic Journalist* [Prague] 31 (October 1984): 13–19; (November 1984): 11–13.

Todd, Judith. "The News Business: Bourgeois Research and the Mass Media." *World Marxist Review* 16 (July 1973): 123–131.

Toward a New Information Order. Special issue. NACLA *Report on the Americas* 16 (July–August 1982).

Tran, Van Dingh. "Non-Alignment and Cultural Imperialism." *Black Scholar* 8 (December 1976): 39–48.

Tuchman, Gaye, ed. *The TV Establishment: Programming for Power and Profit*. Englewood Cliffs, N.J.: Prentice-Hall, 1974.

Wells, Alan. *Picture Tube Imperialism: A Case Study of Television*. Maryknoll, N.Y.: Orbis Books, 1972.

Welty, Gordon. "Communication, Fetishism, and Ideology." In *Studies in Communication*. Vol. 2. Edited by S. Thomas. Norwood, N.J.: Ablex, 1984.

Williams, Raymond. *Communication*. Baltimore: Pelican, 1968.

———. *Marxism and Literature*. Oxford: Oxford University Press, 1977.

———. *Problems in Materialism and Culture*. London: Verso, 1980.

———. *Television: Technology and Cultural Form*. New York: Schocken Books, 1974.

Outer Space, International Broadcasting, and the Direct Broadcast Satellite (DBS)

Abel, Elie, and Philip H. Power. "Third World vs. the Media." *New York Times Magazine*, 21 September 1980, pp. 116–122, 128, 133.

Abshire, David M. *International Broadcasting: A New Dimension of Western Diplomacy*. The Washington Papers, vol. 4, no. 35. Beverly Hills, Calif., and London: Sage, 1976.

American Society of International Law, Panel on International Telecommunications Policy. *The International Telecommunication Union: Issues and Next Steps*. Occasional Paper, no. 10. New York: Carnegie Endowment for International Peace, 1971.

Androunas, Elena, and Y. Zassursky. "Protecting the Sovereignty of Information." *Journal of Communication* 29 (Spring 1979): 186–192.

Aspen Institute. *Control of the Direct Broadcast Satellite: Values in Conflict*. Washington, D.C.: Aspen Institute Publications, 1974.

Aviation Week and Space Technology. Weekly aviation periodical expanded in 1958 to cover space developments.

Beichman, Arnold. "The Story of Radio Free Europe." *National Review*, 2 November 1984, pp. 29–33.

Berrada, A. *Frequencies for Broadcasting Satellites*. International Broadcast Institute Monographs. London: International Broadcast Institute, 1972.

Bogart, Leo. "Projecting America." *Society* 12 (September–October 1975): 57–61.

Bowers, Raymond, Alfred M. Lee, and Cary Hershey, eds. *Communications for a Mobile Society: An Assessment of New Technology*. Beverly Hills, Calif.: Sage, 1978.

Boyd-Barrett, Oliver. *The International News Agencies*. Beverly Hills, Calif.: Sage, 1980.

Branscomb, Anne W. "Waves of the Future: Making WARC Work." *Foreign Policy*, no. 34 (Spring 1979): 139–148.

Brightbill, George D. *Communications and the U.S. Congress: A Selectively Annotated Bibliography of Committee Hearings, 1870–1976*. Washington, D.C.: Broadcast Education Association, 1978.

Brown, Seyom, Nina W. Cornell, Larry L. Fabian, and Edith Brown Riess. *Regimes for the Ocean, Outer Space, and Weather*. Washington, D.C.: Brookings Institution, 1978. (Sponsored by the National Science Foundation, NASA, and the Rockefeller Foundation)

Browne, Donald R. *International Radiobroadcasting: The Limits of the Limitless Medium*. New York: Praeger, 1982.

Butler, Richard E. "World Administrative Radio Conference for Planning Broadcasting Satellite Service." *Journal of Space Law* 5 (Spring and Fall 1977): 93–99.

Bystrom, John. "Peacesat." *World Health* (January 1978): 10–13.

Cherry, Colin. "Thoughts on the Relevance of the 'Communication Explosion' to the Future of World Order." In *Intercommunications Among Peoples*. Edited by Michael H. Prosser. New York: Harper & Row, 1973.

Chronicle of International Communication. Published from 1979 to 1986 by the International Communication Projects, Washington, D.C.

Clarke, A. C. *Voices from the Sky: Previews of the Coming Space Age*. New York: Harper & Row, 1965.

Codding, George A., Jr. "The Politicization of the International Telecommunications Union: Nairobi and After." In *Policy Research in Telecommunications*. Edited by Vincent Mosco. Norwood, N.J.: Ablex, 1984.

———. "Sky Wars: Competition for the Market in American Business Communications Involving Private Links for Voice and Computer Data by Means of Satellites." *The Economist*, 21 April 1979.

Codding, George A., Jr., and Anthony M. Rutkowski. *The International Telecommunication Union in a Changing World*. Dedham, Mass.: Artech House, 1982.

Contreras, James Larson, John K. Mayo, and Peter Spain. *Cross Cultural Broadcasting*. Paris: UNESCO, 1976.

Culture and Communication. Special issue. *Cultures* 6, no. 3 (1979).

d'Arcy, Jean. "Direct Broadcast Satellites and the Right to Communicate." In *The*

Right to Communicate: Collected Papers. Edited by Jim Richstad and L. S. Harms. Honolulu: University of Hawaii Press, 1977.

deCardona, Elizabeth. "American Television in Latin America." In *Mass Media Policies in Changing Cultures*. Edited by George Gerbner. New York: Wiley, 1977.

Democratic Journalist [Prague]. Published since 1950 by the International Organization of Journalists.

Eckhard, Frederic, ed. *Issues 40: Issues Before the 40th General Assembly of the United Nations*. New York: United Nations Association of the United States of America, 1985.

Eek, Hilding A. "Principles Governing the Use of the Mass Media as Defined by the United Nations and Unesco." In *National Sovereignty and International Communication*. Edited by Kaarle Nordenstreng and Herbert I. Schiller. Norwood, N.J.: Ablex, 1979.

Fawcett, J. E. S. "Satellite Broadcasting." *World Today* 27 (February 1971): 76–82.

Fisher, Desmond, and L. S. Harms. *The Right to Communicate: A New Human Right*. Dublin: Boole Press, 1983.

Fisher, Glen. *American Communication in a Global Society*. Norwood, N.J.: Ablex, 1979.

Frederick, Howard H. "*La Guerra Radiofónica:* Radio War Between Cuba and the United States." In *The Critical Communications Review*, Vol. 2. Edited by Vincent Mosco and Janet Wasko. Norwood, N.J.: Ablex, 1984.

Galloway, Jonathan F. *The Politics and Technology of Satellite Communication*. Lexington, Mass.: Lexington Books, 1972.

———. "Worldwide Corporations and International Integration: The Case of Intelsat." *International Organization* 24 (Summer 1970): 503–519.

Global Communications in the Space Age: Toward a New ITU. Report of international conference sponsored by the John and Mary R. Markle Foundation and the Twentieth Century Fund, New York: 1972.

Goldsen, James M., ed. *Outer Space in World Politics*. New York: Praeger, 1963.

Great Soviet Encyclopedia. Vol. 10, S. V. "artificial earth satellites"; Vol. 14, S. V. "space law," "space communications," and "space television." New York: Macmillan, 1973.

Gronberg, Tom, and Kaarle Nordenstreng. "Approaching International Control of Satellite Broadcasting." *Democratic Journalist* [Prague], no. 3 (1974): 12–15.

Hachten, William A. *Muffled Drums: The News Media in Africa*. Iowa City: Iowa State University Press, 1971.

Hamelink, Cees J. *Cultural Autonomy in Global Communications: Planning National Information Policy*. New York: Longman, 1983.

———. *Finance and Information: A Study of Converging Interests*. Norwood, N.J.: Ablex, 1983.

———. *Transnational Data Flows in the Information Age*. Lund, Sweden: Studentlitteratur, 1984.

Harms, L. S., and Jim Richstad. "The Right to Communicate: Status of the Concept." In *Intercultural and International Communication*. Edited by Fred Casmir. Washington, D.C.: University Press of America, 1978.

Harvey, Dodd L., and Linda C. Ciccoritti. *U.S.–Soviet Cooperation in Space*. With foreword by Ambassador Foy D. Kohler. Miami: International Affairs Center for Advanced International Studies, University of Miami, 1974.

Head, Sydney W., ed. *Broadcasting in Africa: A Continental Survey of Radio and Television*. Philadelphia: Temple University Press, 1974.

Heffner, Richard D. "Open Sky or Pre-Censorship: Moscow Against Television by Satellite." *Ost-Europa* 27 (July 1974).

Hochman, Sandra, with Sybil Wong. *Satellite Spies: The Frightening Impact of a New Technology*. Indianapolis: Bobbs-Merrill, 1976.

Horner, Charles. "Outer Space and Earthly Politics: How the Communist and Third World Advocates of a 'New World Information Order' Are Seeking Control of Satellite Communications." *American Spectator* 12 (February 1979): 11–14.

Howell, W. J., Jr. *World Broadcasting in the Age of the Satellite: Comparative Systems, Policies, and Issues in Mass Telecommunication*. Norwood, N.J.: Ablex, 1986.

Humphlett, Patricia E. "Use of the Geostationary Orbit and U.S. Participation in the 1985 World Administrative Radio Conference." Staff report prepared for U.S. Congress. Congressional Research Service. Library of Congress.

Information Economics and Policy. Quarterly. Amsterdam.

Intermedia. [London] International Broadcast Institute Quarterly.

International Telecommunication Union. *Thirteenth Annual Report on Telecommunications and the Peaceful Uses of Outer Space*. Geneva: International Telecommunication Union, 1974.

Jansky, Donald M., and Michel C. Jeruchim. *Communication Satellites in the Geostationary Orbit*. Dedham, Mass.: Artech House, 1983.

Janus, Noreene. "Advertising and the Creation of Global Markets: The Role of the New Communication Technologies." In *The Critical Communications Review*. Vol. 2. Edited by Vincent Mosco and Janet Wasko. Norwood, N.J.: Ablex, 1984.

Kolosov, Y., and V. Kulebyakin. "Television via Outer Space: Some International Problems and Agreements." *New Times* [Moscow], no. 8 (February 1977): 18–19.

LeDuc, Don R. "The DBS Policy Paralysis in Western Europe: Seeking a Solution Among the Symptoms." In *Policy Research in Telecommunications*. Edited by Vincent Mosco. Norwood, N.J.: Ablex, 1984.

Lee, Chin-Chuan. *Media Imperialism Reconsidered: The Homogenizing of Television Culture*. Beverly Hills, Calif.: Sage, 1980.

Levy, S. A. "Intelsat Technology, Politics, and the Transformation of a Regime." *International Organization* 29 (Summer 1975): 655–681.

Lindsay, Robert. "Needed: Ground Rules for the Game in Space." *Media Asia* [Singapore.] 3, no. 1 (1976).

Luyken, Georg-Michael. *Direct Broadcasting Satellites (DBS): An Interdisciplinary Analysis*. Gottingen, Germany, 1984.

McAnany, Emile G. "Television: Mass Communication and Elite Controls." *Society* 12 (September–October 1975): 41–46.

MacBride, Sean et al. *Many Voices, One World: Towards a New, More Just, and More Efficient World Information and Communication Order*. London: Kogan-Page, 1980.

McPhail, Thomas L. *Electronic Colonialism: The Future of International Broadcasting and Communication*. Beverly Hills, Calif.: Sage, 1981.

Maddox, Brenda. *Beyond Babel: New Directions in Communications*. New York: Simon and Schuster, 1972.

Marks, David. "Broadcasting Across the Wall: The Free Flow of Information Between East and West Germany." *Journal of Communication* 33 (Winter 1983): 46–55.

Marks, Leonard. "Politics of World Information." *Society* 12 (September–October 1975): 26–29.

Martin, Thomas H., Richard B. Byrne, and Dan J. Wedemeyer. "Balance: An Aspect of the Right to Communicate." *Journal of Communication* 27 (Spring 1977): 158–162.

The Media Crisis . . . a Continuing Challenge. World Press Freedom Committee's Rex Rand Fund, Washington, D.C.: (Contains text of Talloires Declaration), 1982.

Michaelis, Anthony R. *From Semaphore to Satellite*. Geneva: International Telecommunication Union, 1965.

Mickelson, Sig. *America's Other Voice: The Story of Radio Free Europe and Radio Liberty*. New York: Praeger, 1983.

———. "Communications by Satellites." *Foreign Affairs* 68 (October 1969): 67–79.

Mosco, Vincent, ed. *Policy Research in Telecommunications*. Norwood, N.J.: Ablex, 1984.

Mowlana, Hamid. *Global Information and World Communication: New Frontiers in International Relations*. White Plains, N.Y.: Longman, 1986.

———. "Political and Social Implications of Communications Satellite Applications in Developed and Developing Countries." In *Communications Yearbook I*. Edited by Brent D. Rubin. New Brunswick, N.J.: Transaction Books, 1977.

Musolf, Lloyd D., ed. *Communications Satellites in Political Orbit*. San Francisco: Chandler, 1968.

Naesselund, Gunnar. "International Problems of TV via Satellite: A Major Debate at UNESCO's Recent General Conference." *Unesco Courier* (February 1973): 21–23.

———. "Relations Between and Perspectives Within 'Development Support Communication,' 'Communication Policy Research and Planning,' and 'The Right to Communicate' as seen by UNESCO." In *The Right to Communicate: Collected Papers*. Edited by Jim Richstad and L. S. Harms. Honolulu: University of Hawaii Press, 1977.

Network Project. *The Case Against Satellites*. Pamphlet, no. 7. New York: The Network Project. Spring, 1974.

Newsom, Carroll V. "Communication Satellites: A New Hazard to World Cultures." *Educational Broadcasting Review* 7 (April 1973).

Nordenstreng, Kaarle, with Lauri Hannikainen. *The Mass Media Declaration of Unesco*. Norwood, N.J.: Ablex, 1984.

Nordenstreng, Kaarle, and Herbert I. Schiller, eds. *National Sovereignty and International Communication*. Norwood, N.J.: Ablex, 1979.

O'Brien, Rita Cruise. *Information, Economics, and Power: The North–South Dimension*. Boulder, Colo.: Westview Press, 1983.

"The Politics of Information." Interview with Stan Swinton, director of world services for the Associated Press. *Multinational Monitor* 1 (December 1980): 12–14.

Pelton, Joseph N. *Global Talk*. Alphen aan den Rijn, Netherlands: Sijthoff and Noordhoff, 1981.

Ploman, Edward W. *Space, Earth, and Communication*. Westport, Conn.: Quorum Books, 1984.

Pomorski, Jerzy Mikulowski. "Right to Communicate: National Legislation or International Policy?" *The Right to Communicate: Collected Papers*. Edited by Jim Richstad and L. S. Harms. Honolulu: University of Hawaii Press, 1977.

Pool, Ithiel de Sola. "The Changing Flow of Television." *Journal of Communication* 27 (Spring 1977): 139–149.

———. "Direct Broadcast Satellites and Cultural Integrity." *Society* 12 (September–October 1975): 47–56.

Powell, Jon T. *International Broadcasting by Satellite: Issues of Regulation, Barriers to Communication*. Westport, Conn.: Quorum Books, 1985.

Prentiss, Stan. *Satellite Communications*. Blue Ridge Summit, Pa.: TAB Books, 1983.

Presidential Study Commission on International Radio Broadcasting. *Study of Radio Liberty and Radio Free Europe with Recommendations for their Future*. Washington, D.C.: Government Printing Office, 1973.

Prosser, Michael H., ed. *Intercommunications Among Peoples*. New York: Harper & Row, 1973.

Queeney, Kathryn M. *Direct Broadcast Satellites and the United Nations*. Alphen aan den Rijn, Netherlands: Sijthoff and Noordhoff, 1978.

Quester, George H. "Transboundary Television." *Problems of Communism* 23 (September–October 1984): 76–87.

Read, William H. "Global TV Flow: Another Look." *Journal of Communication* 26 (Summer 1976): 69–73.

Review of International Broadcasting. Published monthly since 1976, Box 1684, Enid, Oklahoma.

Richstad, Jim, and L. S. Harms, eds. *The Right to Communicate: Collected Papers*. Honolulu: University of Hawaii Press, 1977.

Riegel, Oscar W. "Satellite Communication and National Power." In *Mass Media Policies in Changing Cultures*. Edited by George Gerbner. New York: Wiley, 1977.

Righter, Rosemary. "Battle of the Bias." *Foreign Policy,* no. 34 (Spring 1979): 121–138.

Robinson, Glen O., ed. *Communications for Tomorrow*. Report of Aspen Institute Special Task Force on Communications Policymaking. New York: Praeger, 1978.

Ronalds, Francis S., Jr., "Voices of America." *Foreign Policy,* no. 34 (Spring 1979): 154–160.

"Saga of Super Sound: Act IV Scene 1." *Chronicle of International Communication* 5 (September 1984): 4–6.

Sarnoff, David. "America's Mirror Image." Speech given on June 14, 1962. New York: National Broadcasting Company. n.d.

Sarnoff, Robert W. "Freedom to Communicate." Speech given on February 9, 1973. New York: RCA. n.d.

———. "Satellite Communications: Concepts for a New Age." Speech given on October 2, 1975. New York, RCA, n.d.

Satellite Broadcasting and Communications Policy: DBS Comes of Age. Special issue. *Journal of Communication* 30 (Spring 1980).

Schauer, William. *The Politics of Space: A Comparison of the Soviet and American Space Program*. New York: Holmes and Meier, 1976.

Schiller, Dan. *Telematics and Government*. Norwood, N.J.: Ablex, 1982.

Shternfeld, Ari. *Soviet Space Science*. Translated by the Technical Documents Liaison Office, Wright Patterson Air Force Base, Ohio. New York: Basic Books, 1959.

Signitzer, Benno. *Regulation of Direct Broadcasting from Satellites—The UN Involvement*. New York: Praeger, 1976.

Silk, Leonard. "The Impact on the American Economy." In *Outer Space: Prospects for Man and Society*. Edited by Lincoln P. Bloomfield. Englewood Cliffs, N.J.: Prentice-Hall, for the American Assembly, Columbia University, 1962.

Singh, K., and B. Gross. "'MacBride': The Report and the Response. The Press, the U.S., and UNESCO." *Journal of Communication* 31 (Autumn 1981): 104–117.

Smith, Delbert D. *Communication via Satellite: A Vision in Retrospect*. Alphen aan den Rijn, Netherlands: Sitjhoff-Noordhoff, 1976.

———. *Teleservices via Satellite: Experiments and Future Perspectives*. Boston: Sijthoff and Noordhoff, 1978.

Smith, Don D. "Mass Communications and International Image Change." *Journal of Conflict Resolution* 17 (March 1973): 115–129.

Snow, Marcellus S. *International Commercial Satellite Communications: Economic and Political Issues of the First Decade of Intelsat*. New York: Praeger, 1976.

Space Communication and Broadcasting—An International Journal [Netherlands]. Published quarterly since 1983.

"Space Technology and Africa." *Africa* (December 1977): 50–57.

Stanton, Frank. "Will They Stop Our Satellites?" *New York Times*, 22 October 1972, sec. 2, p. 39.

Sterling, Christopher. *International Telecommunications and Information Policy*. Washington, D.C.: Communications Press, 1984.

Teleclippings. Periodical reprints from world press. Public Relations Division of General Secretariat, International Telecommunication Union, Geneva.

Telecommunication Journal [Geneva]. Periodical published by the International Telecommunication Union.

Telecommunications Development: The U.S. Effort—Authorities Review Post–UNESCO Policies. Special issue. *Journal of Communication* 35 (Spring 1985).

Telecommunications Policy. Quarterly. London.

Tobin Foundation. *The Structural Issues in Global Communications*. Washington, D.C.: Tobin Foundation, 1982. A report based on a meeting at Leeds Castle, Kent, England, of twenty-five world leaders from broadcasting, telecommunications, manufacturing, and government.

Transnational Data and Communications Report. Monthly. Washington, D.C.

Twentieth Century Fund. *The Future of Satellite Communications. Resource Management and the Needs of Nations*. Second Report of the Twentieth Century Fund Task Force on International Satellite Communications. New York: Twentieth Century Fund, 1970.

UNESCO. *Broadcasting from Space*. Reports and Papers on Mass Communications, no. 60. Paris: UNESCO, 1970.

———. *Communication Satellites for Education, Science, and Culture*. Reports and Papers on Mass Communications. no. 53. Paris: UNESCO, 1967.

———. *Communication in the Space Age: The Use of Satellites by the Mass Media*. Paris: UNESCO, 1968.

———. *Cross-cultural broadcasting*. Reports and Papers on Mass Communications, no. 77. Paris: UNESCO, 1976.

———. *External Radio Broadcasting and International Understanding*. Reports and Papers on Mass Communications, no. 81, Paris: UNESCO, 1977.

———. *A Guide to Satellite Communication*. Reports and Papers on Mass Communications, no. 66. Paris: UNESCO, 1972.

———. *New Values and Principles of Cross-cultural Communication*. Reports and Papers on Mass Communications, no. 85. Paris: UNESCO, 1979.

———. *The Right to Communicate: A Status Report*. Reports and Papers on Mass Communications, no. 94. Paris: UNESCO, 1981.

———. *Television Traffic—A One-Way Street?* Reports and Papers on Mass Communications, no. 70. Paris: UNESCO, 1974.

———. *Towards Realistic Communication Policies: Recent Trends*. Reports and Papers on Mass Communications, no. 76. Paris: UNESCO, 1976.

———. *Transnational Communication and Mass Media Industries*. Reports and Papers on Mass Communications, no. 92. Paris: UNESCO, 1980.

United Nations. Reports of the Committee on the Peaceful Uses of Outer Space since 1968, including reports from the COPUOS Legal Subcommittee and other subcommittees working on DBS. New York: United Nations.

United States Advisory Commission on Public Diplomacy. Annual reports. Washington, D.C.: United States Advisory Commission on Public Diplomacy, 1985, 1986.

U.S. Central Intelligence Agency, Joint Publications Research Service. *Worldwide Report: Telecommunications Policy, Research and Development*.

U.S. Congress. House. Committee on Foreign Affairs. Subcommittee on International Relations. *Modern Communication and Foreign Policy*. Report no. 5, with Part X of *Winning the Cold War: The U.S. Ideological Offensive: Hearings*. 90th Cong., 1st sess., 1967.

U.S. Congress. House. Committee on Foreign Affairs. Subcommittee on National Security Policy and Scientific Developments. *Satellite Broadcasting: Implications for Foreign Policy*. 91st Cong., 1st sess., 1969.

U.S. Congress. House. Committee on Government Operations. *International Shortwave Broadcasting and Direct Broadcast Satellites: VOA, RFE/RL, Radio Martí: Twentieth Report*. 97th Cong., 1st sess., 1981.

U.S. Congress. House. Committee on Government Operations. Subcommittee on Government Information and Individual Rights. *International Broadcasting: Direct Broadcast Satellites: Hearings*. 97th cong., 1st sess., 1981.

U.S. Congress. House. Committee on Government Operations. Subcommittee on Military Operations. *Government Use of Satellite Communications: Hearings*. 89th Cong., 2d sess., 1966.

U.S. Congress. House. Committee on Science and Technology. *Opportunities for International Cooperation in Space as well as the Economic and Technological Implications Inherent in such Cooperation*. 95th Cong., 2d sess., 1978.

U.S. Congress. House. Committee on Science and Astronautics. *Report on the International Conference of States on the Distribution of Program-carrying Signals Transmitted by Satellite, in Brussels*. 93rd Cong., 2d sess., 1974.

U.S. Congress. House. Committee on Science and Technology. Subcommittee on Space Science and Applications. *International Space Law*. 94th Cong., 2d sess., 1976.

U.S. Congress. Senate. Committee on Aeronauticaal and Space Sciences. *Communication Satellites: Technical, Economic, and International Developments*. Staff report. 87th Cong., 2d sess., 1962.

U.S. Congress. Senate. Committee on Aeronautical and Space Sciences. *International Cooperation in Outer Space: A Symposium*. Edited by Eilene Galloway. 92nd Cong., 1st sess., 1971.

U.S. Congress. Senate. Committee on Aeronautical and Space Sciences. *International Cooperation and Organization for Outer Space*. Staff report. 89th Cong., 1st sess., 1965.

U.S. Congress. Senate. Committee on Aeronautical and Space Sciences. *Soviet Space Programs, 1971–75: Goals and Purposes, Organization, Resource Allocations, Attitudes Toward International Cooperation and Space Law*. 94th Cong., 2d sess., 1976.

U.S. Congress. Senate. Committee on Foreign Relations. Subcommittee on International Operations. *The Implications of International Communications and Information: Hearings*. 95th Cong., 1st sess., 1976.

U.S. General Accounting Office. *Improvements Made, Some Still Needed in Management of Radio Free Europe/Radio Liberty*. Report, 1981.

U.S. Information Agency. *Wireless File*, January–March, 1985. Daily compilation by USIA of articles, speeches, and government news releases; telexed to United States embassies abroad. Articles on the United States Information Agency, Voice of America, Worldnet, U.S. activities in space.

"United States Proposes Voluntary Principles on Direct Broadcasting by Satellite." *Department of State Bulletin*, 22 April 1974.

"USA Participation in the International Telecommunications Union: A Study of Policy Alternatives." Prepared for the State Department. Washington, D.C.: Center for Strategic and International Studies, Georgetown University, 1985.

Valdez, Armando. "Development of New Telecommunication Resources in a Deregulated Enironment: an Examination of FCC Policy on Direct Broadcast Satellites." *Policy Research in Telecommunications*. Edited by Vincent Mosco. Norwood, N.J.: Ablex, 1984.

Varis, Tapio. "Global Traffic in Television." *Journal of Communication* 24 (1974): 102–109.

Weil, Gordon. *Communicating by Satellite: An International Discussion*. New York: Twentieth Century Fund/Markle Foundation, 1969. A report of a conference of experts meeting at Talloires, France, September 21–25, 1969.

White, Curtis T. "Uprooting the Squatters." *Foreign Policy*, no. 34 (Spring 1979): 148–153.

Wigand, Rolf T. et al. "Transborder Data Flow, Informatics, and National Policies." *Journal of Communication* 34 (Winter 1984): 153–175.

Wilcox, Dennis L. *Mass Media in Black Africa: Philosophy and Control*. New York: Praeger, 1975.

Zasursky, Y. N. "'Free Flow of Information': The Cold War and Reducing International Tension." *Democratic Journalist* [Prague], no. 3 (1974): 7–11.

Ziebarth, E. W. "The Electronic Media as Instrument of Social Control in Peoples' Republic of China." 1976. Mimeo.

Index